Assetization

Inside Technology

Edited by Wiebe E. Bijker, W. Bernard Carlson, and Trevor Pinch

A list of titles in this series appears at the back of the book.

Assetization

Turning Things into Assets in Technoscientific Capitalism

Edited by Kean Birch and Fabian Muniesa

The MIT Press
Cambridge, Massachusetts
London, England

The open access edition of this book was made possible by generous funding from Arcadia—a charitable fund of Lisbet Rausing and Peter Baldwin.

ARCADIA
A charitable fund of Lisbet Rausing and Peter Baldwin

This book was set in Stone Serif and Stone Sans by Westchester Publishing Services.

Library of Congress Cataloging-in-Publication Data

Names: Birch, Kean, editor.
Title: Assetization : turning things into assets in technoscientific capitalism / edited by Kean Birch and Fabian Muniesa.
Description: Cambridge, Massachusetts : The MIT Press, [2020] | Series: Inside technology | Includes bibliographical references and index.
Identifiers: LCCN 2019047640 | ISBN 9780262539173 (paperback)
Subjects: LCSH: Social capital (Sociology) | Commodification. | Technology--Social aspects. | Temporary employment. | New business enterprises. | Uncertainty. | Capitalism.
Classification: LCC HM708 .A78 2020 | DDC 302--dc23
LC record available at https://lccn.loc.gov/2019047640

150377129

Contents

Acknowledgments

This book started out with a series of panels at the 2016 joint Society for the Social Studies of Science and European Association for the Study of Science and Technology Conference in Barcelona, Spain. We extend our gratitude to the 4S/EASST 2016 Conference organizing committee: Nerea Calvillo, Miquel Domènech, Daniel López, Vincenzo Pavone, Israel Rodríguez-Giralt, Carmen Romero-Bachiller, Tomás Sánchez Criado, and Francisco Tirado. Since then, our intellectual project has involved a number of twists and turns, and we are thankful to all those people who have contributed to the successful completion of the book, including all the contributors to the conference panels at the start of the project; the series editors of Inside Technology, Trevor Pinch and Wiebe Bijker, for their support and feedback; the anonymous proposal and manuscript reviewers for their excellent suggestions; the various colleagues who have provided each of us with their invaluable comments and thoughts on our draft chapters; and, especially, Katie Helke for steering us so assuredly and thoughtfully through the editorial process. More specific thanks to Sara Angeli Aguiton, Laurent Beduneau-Wang, Ronan Bolton, Daniel Breslau, Lise Cornilleau, Dror Etzion, Bernard Forgues, David Hawkey, Luka Jakelja, Charlotte Kroløkke, Emmanuel Kypraios, Shimrit Lee, Marcela Linkova, Michaela Mayrhofer, Andrea Mennicken, Yuval Millo, Shelly Ronen, Stephen Timmons, Paraskevas Vezyridis, Janette Webb, and Bernhard Wieser.

1 Introduction: Assetization and Technoscientific Capitalism

Kean Birch and Fabian Muniesa

Introduction

Contemporary capitalism is increasingly defined by its technoscientific aspects—that is, by the development of new technological products and services (e.g., smartphones, apps, platforms), the emergence of trendy scientific specialties (e.g., big data, AI, biotech, fintech), and the alignment of innovation processes, actors, and institutions with powerful investment rationalities and financial imperatives. The very notion of "technoscientific capitalism"—terminology whose origins can be traced back at least to Jean-Francois Lyotard (1984, 1992)—aptly captures the defining problematics of our political-economic time.

But how can we best characterize the gist of this moment? What is the prevailing form that things take within technoscientific capitalism? Because of its innovation-based nature and its concomitant promissory rationale, technoscientific capitalism has often been interpreted in terms of a speculative logic (e.g., Sunder Rajan 2006, 2017). The prevalence of processes of commercialization and privatization have repeatedly motivated a reading of technoscientific capitalism as a commodification movement that orients science and technology toward a market destiny (e.g., Mirowski 2011; Berman 2012). It is undeniable that the market plays a crucial role in technoscientific capitalism, offering the commodity form as a template of choice for all things scientific and technological, and prompting accordingly the relevance of a speculative gaze. The full picture, though, lacks analytic traction if we rely solely on this perspective.

Our argument in this book is that the dominant *form* that technoscientific capitalism affords is not the commodity but the asset, and that the financial contours it entails are not those of market speculation but of capital investment. Those are very different things, and the aim of contributors to this book is to flesh out this central point. Our perspective, then, is that we cannot characterize and analyze technoscientific capitalism solely in commodity terms anymore, not in the era of Uber and Airbnb, Google and Amgen. Can we understand the "unicorns" stalking Silicon Valley (i.e., firms whose notional valuations top $1 billion, promising huge returns to shrewd investors), the biopharmaceutical firms ramping up drug prices (while buying back their shares to shore up their share value), and governments turning the atmosphere into financial assets (and giving them away to polluters) without struggling within prevailing conceptions of the commercialization, marketization, and commodification of technoscientific research and innovation? As the latest stage of capitalism entrenches, it becomes even more difficult to see what is driving the accumulation of capital. As the marginal cost of production approaches zero, it is increasingly evident that productivity increases will not lead to sustainable profits. What will? Business model sorcery perhaps?

Intriguing things are going on indeed, we contend, as a consequence of an emerging "asset form" that has come to replace the commodity as the primary basis of contemporary capitalism. By asset, we mean something that can be owned or controlled, traded, and capitalized as a revenue stream, often involving the valuation of discounted future earnings in the present—it could be a piece of land, a skill or experience, a sum of money, a bodily function or affective personality, a life-form, a patent or copyright, and so on. Discounting reflects an assumption about the future value of money, which is framed by expectations about future inflation, risks, and uncertainties (Muniesa et al. 2017). Assets can be bought and sold, yes. But the point is to get a durable economic rent from them, not to sell them in the market today; here we use the term *rent* to mean the extraction of value through the ownership and control of an asset, which usually entails limiting access to it (Birch 2020).

How do things become assets, then? They are made so: the asset form is not, it is important to stress, the consequence of some inherent or embodied quality. One intention with this book is to show how assets are constructed, how a variety of things are and can be turned into assets, examining the interests, activities, skills, organizations, and relations entangled in this process. Another is to stress that technoscientific capitalism entails specific practices that make the uncertainty inherent in innovation understandable and calculable as part of a broader capitalist system. The asset form reflects the tumult in contemporary technoscientific capitalism, in which it becomes harder and harder to draw clear boundaries around what counts as or comes to constitute capitalism. For example, it is not clear whether the entitlements created by governments (e.g., emissions credits, see Felli 2014) or the expansion of private ownership claims over more and more aspects of our lives (e.g., social media content, see Arvidsson and Colleoni 2012) reflect the dynamics of earlier, entrepreneurial capitalist processes. Are they something new altogether? It is for these and comparable reasons that we think it is now important to turn our gaze toward the specificity of the asset form and to the "assetization" process as key political-economic practices (Birch 2017a; Muniesa et al. 2017).

What Is the Asset Form? What Is Assetization?

A simple question like "What is an asset?" is hard to answer without resorting to an understanding of the asset form embedded in a range of epistemic and ontological assumptions. To an accountant or businessperson, the term *asset* might have a particular resonance with a balance sheet of liabilities and equity. To a Marxist political economist, the term might hold no relevance, with a preference for the term *capital* instead. To a banker or a financial analyst, the term might mean the securities in an investment portfolio and prompt ideas of valuation, hedging, and arbitrage. And to the middle-class homeowner, it might refer to a way to secure a family's future. As we want to make evident here, there are many different orientations that people can have toward the definitions of asset and assetization. At its base, though, an asset is defined technically by the International Accounting Standards Board (IASB) as "a resource controlled by the entity as a result of past events and from which future economic benefits are expected to flow to the entity" (Burton and Jermakowicz 2015, 39). We can already see how ductile this terminology is, however, reflecting many conceptual uses that do not fall within the scope of the technical jargon of accounting standards.

The notion of asset certainly speaks to the notion of capital, both in the vernaculars of financial accounting and business life and in the formalization of accountable profit within capitalism (Levy 2014). Something whose "control" may warrant "future economic benefits" pretty much sounds like capital—that is, wealth (money or something else) considered in terms of an investment with an aim to generate revenues, yields, or rents in a more or less distant future. Both notions—capital and asset—often appear in the literature as substantive things (something that someone may or may not have), but they are also open to interpretations in terms of form, process, condition, or relation (Levy 2017). For us, the term *asset* is less loaded with theoretical controversies and terminological quandaries than the term *capital* (e.g., Hodgson 2014). And this is the main, practical reason that explains our inclination for the former in the pages that follow. Another obvious reason is to provide our inquiry with an opportunity to retain the processual intuition of the asset not as a thing but as a form—it may be too late to impose this standpoint on the notion of capital.

Our adoption of the notion of the asset is also meant to disentangle considerations of the asset as both an objective resource (aka factor of production) and as a subjective value (or effect of valuation practices) (Muniesa 2012)—that is, its form *and* the condition it engenders. The term *capitalization* can certainly work usefully in the direction of capturing the processual aspect of the asset form (Leyshon and Thrift 2007). That being said, unless explicitly defined as a wide cultural process consisting precisely in turning things into assets (e.g., Muniesa et al. 2017), that term runs the risk of directing attention to its special meanings in the accountant's technical terminology. Using *assetization* here serves our purpose better, since it emphasizes the socially transformative character of the phenomenon of turning things into assets—it can refer to that phenomenon proper as much as to its societal consequences (i.e., "asset condition").

The notion of capital, in contrast, is used to refer to everything from fixed capital (e.g., machinery) through circulating capital (e.g., goods, labor) to commodity-capital and money-capital in critical political economy (Chiapello 2007). Its underpinning in a theory of labor value, moreover, does not do enough to help us understand contemporary capitalist phenomenon (Nitzan and Bichler 2009). As Durand (2017) argues, even the likes of Thomas Piketty (2014) in his magisterial *Capital in the 21st Century* ends up using the term *capital* in contradictory ways, mixing up the analytical (e.g., capital as factor of production) and the empirical (e.g., capital as wealth). Our focus on the asset form and the process of assetization provides a means for cutting across such potential impasses while enabling overlapping analyses of accounting practices, cultural metaphors, and political-economic trends, each of which entails a different analytical tone.

The notion of assetization surely speaks to the notion of financialization too. Despite its limits and ambiguities, financialization has opened up multiple fruitful paths to characterize contemporary technoscientific capitalism and its wider economic, political, and social contexts (van der Zwan 2014; Christophers 2015). Whether financialization refers to the rise of shareholder value, or the liberalization of financial regulation, or the spread of specific techniques of securitization, or wider cultural transformations involving the emergence of new forms of subjectivity, the idea of financialization certainly captures a movement that involves or implies the turning of things into assets we are concerned with here. The political and policy turn toward finance since the 1970s in the United States, for example, is definitely accompanied by processes of assetization (Krippner 2011). The same applies to transformations of corporate strategy through the spread of financial views and metrics, as the requirements of shareholder value prompt corporate focus on the asset form (Froud et al. 2006).

The gradual realization of the relevance of the asset base in the everyday experiences of personal finance constitutes a prominent feature of these financialization debates (Langley 2008). And the spirit of the resulting asset condition is unquestionably a crucial ingredient of the cultural complex that goes by that name (Aitken 2007). Is there a reason, then, why we prefer *assetization* to *financialization* in the pages that follow? As the motto of the studies gathered in this book suggests—"turning things into assets"—a focus on the asset form calls for an inquiry into the contours this form ought to consist of in specific circumstances. Talking of "turning things into finance" would grant too wide a scope, as finance—and financialization—is a process of more abstract proportions. Besides, it is perfectly possible that processes of assetization can occur outside the professional perimeters usually attributed to the rise of finance and financialization.

What then sets the asset form and assetization process apart? Drawing on the work of Birch and Tyfield (2013) and Birch (2017a, 2017b)—among others—it is possible to identify at least seven aspects that mark out the asset form as a distinct analytical and practical entity worthy of our intellectual interest, and the contributors highlight several further issues in the following pages. First, assets are legal constructs, in that ownership and control rest on the state enforcement of property and control rights (Pistor 2013, 2019; Dreyfuss and Frankel 2015). Critically, assets can entail the separation of rights from the thing involved as well as the differentiation between the ownership and control of an asset through forms of contract and property law. For example, although they cannot be owned, customers can still be classified as an asset in that access to them can be controlled (MacKenzie 2009). Second, assets involve distinct *modes* of ownership and control (Perzanowski and Schultz 2016; Birch 2020). In particular, knowledge, creative, or data assets (e.g., intellectual property rights or IPRs) have been legally instituted to give owners both exclusion rights to the use of the asset itself and to the use of any copies derived from the asset (Frase 2016). Such "flow-through" rights enable asset holders to control the way that copies or derivatives of an asset are used or experienced, in contrast to commodities. For example, no one can tell you how to wear a coat you have bought, but they can tell you how to use the copyrighted music or data they have sold you.

Third, assets often involve forms of "rentiership" in which monopoly control—derived from, for example, IPRs or government fiat—enables the extraction of economic rents (Fuller 2002; Zeller 2008; Birch 2017a, 2020; Mazzucato 2018; Fields 2019). Assets are often unique, meaning that their value derives from their asset specificity (Teece 1986); as such, it is not possible to reproduce them easily, cheaply, or even at all. For example, a parcel of land is unique and nonreproducible, while a specific copyright cannot be legally reproduced without permission. Fourth, as the result of being unique or constructed monopolies, assets have distinct supply and demand logics in which rising asset prices do not lead to new producers or creators entering a market and thereby lowering prices (Birch and Tyfield 2013). Fifth, asset value can be discounted in light of forward-looking expectations about future returns on investments, whether or not those expectations are met. For example, Birch (2015), Cooper (2017), Kiely (2018), and Adkins et al. (2019) highlight the ways that expectations about ever-rising house prices in Anglo-American economies have instituted a new accumulation regime (asset-based economy), which configures social relations and inequalities in particular ways. This raises serious questions about whose future interests come to shape current policies and institutions.

Sixth, asset prices and valuations are subject to the actions of owners who may seek to reduce the economic value of their assets, or turn an asset into another form, transfer ownership, or use it to attract new partners (Lezaun and Montgomery 2015). As this would imply, valuation does not reflect an inherent or fundamental quality (such as labor-power). Rather, value is very much temporal since it changes over time as the result of the configuration of its social, discursive, and material dimensions and uses (Muniesa et al. 2017). In contrast, a commodity's economic value is determined at the specific point of exchange. Finally, as Birch (2017a) argues elsewhere and the previous discussion highlights, asset value and valuation are dynamic, in that they are constituted by an active and ongoing management of that value by social actors who are both internal (e.g., managers) and external (e.g., financial analysts) to an organization. For example, an asset's value is configured by an ecosystem of diverse financial, technoscientific, political, and social actors, ranging from corporate lawyers through stockbrokers and stock analysts to policy-makers and universities (Birch 2016; Owen and Hopkins 2016; Muniesa et al. 2017).

Lineages of the Asset Form in Economics

Important clues to the theorization of the asset form can certainly be gathered from economic thought, particularly from thinkers who have emphasized the institutional nature of this form. Writing at the start of the twentieth century, Thorstein Veblen represents a critical early influence on the subject matter (see Nitzan and Bichler 2009). Veblen (1908a, 1908b) published two journal articles called "On the Nature of Capital" in the *Quarterly Journal of Economics*. Although less well known than his other work, these articles provide an important groundwork for later thinkers, especially in the critical, constructivist tradition (see below). As Veblen understood the term, *asset* is a "pecuniary concept" that "connotes ownership as well as value" (1908b, 518n1), where the latter represents "capitalizable value" (1908a, 105). As such, an asset is tied to pecuniary investment and pecuniary gain in both value and ownership—that is, an asset is not a resource contributing inherently or by itself to industrial efficiency and productivity. An asset is instead a financial entity as much as it is a resource—material or immaterial—that contributes to production; for Veblen, "assets are a matter of capitalization, which is a special case of valuation" (1908a, 121). As capitalized property, an asset reflects the assessment of future earnings that accrue to the owner, rather than rising productivity. Moreover, intangible assets are a way to capitalize "habits of life" like trust and loyalty (e.g., brands), taste (e.g., fashion), love (e.g., gifts), and so on.

John R. Commons features prominently, next to Veblen, in the genealogy of an institutionalist approach to the economics of the asset form. In his *Legal Foundations of Capitalism* (1924), he provided a characterization of the notion of asset as a pivotal piece within a conception of economic value in which the latter would derive not from the properties of things or their commercial appraisal but from their expected earning power. The idea of earning power as the key to valuation was already well established in the incipient curriculum of North American business schools in the interwar period (Dewing 1920). The prime quality of the finance-savvy businessperson ought to reside in the capacity to detect earning power, prospectively, in any possible business venture. Authors such as Commons offered a refined conceptual formulation; the very notion of capital—a "business notion of capital"—had to boil down to the power that assets had by virtue of the return that could be expected from them.

But the seminal economic theorist, duly acknowledged by Commons, on the asset form and condition as a key to the understanding of capital is certainly the neoclassical economist Irving Fisher. The basis of the value of capital, as Fisher made explicit in *The Nature of Capital and Income* (1906) and in *The Rate of Interest* (1907), is the *future*; the value of capital is the present worth of the future income obtained from that capital. The formalization of the rate at which the latter will be discounted in order to obtain the former provides the answer to the prime enigma of business valuation—namely, how to value the risk incurred by the capitalist that puts money in the business. The practice of the allocation of money finds there the fundamental principle that will maximize earning power.

The elaboration of a neoclassical theory of the asset by Fisher reflected, to some extent, the institutional—if not pragmatist—intuitions developed by Veblen or Commons but also later the ideas of John Maynard Keynes. In *The General Theory of Employment, Interest and Money* (1936), Keynes famously offers a conjecture on how such rates form in practice as conventions that prompt the "inducement to invest." Contention existed among economists of the time on how spontaneous or legitimate such an order of value was and on how it ought to be better formalized. But the asset form and condition govern the discussion, and the general approach is to locate within the perspective of an investor the template for the valuation of all things economic.

The quantitative turn in post–World War II financial economics followed, which further complicated the concept's tortuous history (see MacKenzie 2006, 2009). This turn can be read as a further formalization—and mathematical articulation—of the principles and doctrines that configured the economic theorization of asset valuation developed in earlier decades, particularly in the North American landscape (see Bryer 2012, 2013a, 2013b). A key contribution in that respect is the formalization, by Franco Modigliani and Merton Miller, of the conditions in which the cost of capital (or the return required by an investor) can be optimally calculated in order to guide investment policies. Further developments in corporate finance by authors such as Harry Markowitz and William F. Sharpe provided a more complete formalization of the construction of investment portfolios that would minimize risk. Such is the role of asset valuation formulas like the capital asset pricing model (CAPM). A further landmark in that history is the development by Fischer Black, Myron Scholes, and Robert C. Merton of a mathematical model (the Black-Scholes model) that provided criteria for determining the supposedly fundamental underlying prices of derivative contracts (financial options and futures) and which was later abundantly used in order to hedge portfolios of financial assets—that is, to cancel their risk (Bjerg 2014).

Such developments were both prompted by and operational in the augmentation and complexification of North American financial markets from the 1960s onward. These offered a growing variety of instruments—portfolios of bonds, stocks, futures, and options—from which to seek earning power and on which to exert financial imagination. Developments in the area of financial accounting and business valuation proper were perhaps less salient and basically revolved around the refining of the recipes already found in Fisher. Notions such as discounted cash flow (DCF), net present value (NPV), and weighted average cost of capital (WACC) offer the kernel of the practices and techniques of capital budgeting, investment policy, and financial analysis found worldwide in businesses of all kinds (Doganova 2018a, 2018b). Key to the understanding of the history and practice of such calculative instruments is their concomitant association with distinct managerial practices and epistemic visions. The entire doctrine of the shareholder value maximization and the refinement of precepts for corporate governance strategy by financial economists such as Michael C. Jensen and Eugene Fama is a case in point, as such visions aim at transforming the very nature of the corporation (Baars and Spicer 2017).

As the above should illustrate, there has been and continues to be a long-standing interest in neoclassical circles in assets and asset valuations from a practical (or even practitioner) perspective. Specific examples might include the business interest in and research on reputation as an asset (e.g., Helm et al. 2011), although a more general and pertinent analytical example is the notion of asset specificity introduced by the new institutional economist Oliver Williamson in the 1970s. According to Riordan and Williamson (1985), transaction costs are differentiated by asset specificity, by which they mean that the more an organization's assets (e.g., land, workers, reputation, machinery) are distinct and unique (i.e., specific), the more that "exchange relations take on a progressively stronger bilateral trading character" (367). Simply put, firms need to trade when they are different from one another, but market competition ends up being replaced by other forms of economic governance due to fears about opportunism (since each side does not know the value of the other side's assets) (see also Williamson 1979). Such analytical claims can be traced back to work by Ronald Coase (1937) on the theory of the firm, and they foreground later debates around incomplete contracts that were deemed to result from the uncertainty engendered by asset specificity foreclosing the capacity of business to write affordable complete contracts (see Joskow 1988).

As part of a broader theoretical lineage, concepts like asset specificity have proved highly influential in management circles, especially to academic theories like the resource-based theory of the firm—first proposed by Edith Penrose in the 1950s—and the later capability-based theory of the firm put forward by David Teece and others (see, for example, Pisano 1991; Pitelis and Teece 2009). In both cases, management thinkers theorized organizational performance as the result of the asset base of firms, rather than market competitiveness. As with financial economics, these theoretical concerns in management literatures often collapse into business practices. For example, David Teece founded a consultancy firm—Berkeley Research Group—whose primary service seems to be providing expert economic advice to clients who want to overcome antitrust suits against them. More generally they illustrate the extent to which economic and management scholars, managers and executives, financiers and investors, and others at the sharp edge of business are all concerned with understanding assets as objects of political-economic study and political-economic intervention. Study and intervention feed off each other, in fact, reflecting the influence of abstract models and theories on business decision-making. The capital asset pricing model (CAPM), developed in financial economics, along with the efficient market hypothesis (EMH) came to inform how value was understood *and* created in investor decision-making (Fourcade and Khurana 2017). It is these sorts of relations—between the theory and the practice of asset valuation—that have engendered a growing interest in assets and assetization in critical constructivist scholarly circles, to which we turn next.

Critical and Constructivist Takes on the Asset Form

Early work by Veblen, Commons, and others illustrated the extent to which the asset form has been a conceptual and practical focus for several generations of scholars, especially in the quarters of institutional economics. More recent research streams have provided further bases from which to approach the subject matter in a productive manner, offering perhaps tighter connections to the social and political concerns prompted by contemporary technoscientific capitalism. Those streams are recognizable today through a number of heterogeneous disciplinary tags, such as cultural economy, heterodox economics, critical political economy, critical accounting, and science and technology studies (STS).

Cultural Economy and Anthropologies of Finance

Calls for a qualitative, ethnographic appraisal of economic expertise, financial knowledge, and accounting technique have found fertile terrain in cultural studies in the past two decades. Finance features prominently in academic areas labeled "cultural economy" (Du Gay and Pryke 2002; Pryke and Du Gay 2007) or "cultural political economy" (Sum and Jessop 2013; Jessop, Young, and Scherrer 2015). Valuation processes are made sense of there from the perspective of both the moral imaginaries they carry and the political setup they establish. The implications of financialization—especially the spread of financial views and techniques for which the asset form is an essential ingredient—are aptly tackled from such perspective, including as a set of arrangements that permeate everyday life (see, for example, Langley 2008; McFall 2014; Deville 2015; Fridman 2017).

This emergent area is connected to developments in the anthropology of finance and money that see in global finance the key to understanding the forms that reflect the nexus between cultural meaning and social hierarchy (see Maurer 2006; Hart and Ortiz 2014; Hart 2017). A number of ethnographies of valuation practices in large investment banks have in this respect examined the links between representations of worth among analysts, traders, and accountants; the organizational and technological arrangements within which they evolve; and the structural consequences of their operations (e.g., Beunza and Stark 2004; Zaloom 2006; Ho 2009; Lépinay 2011; Ortiz 2014). Ortiz (2013, 2014), in particular, has introduced the "limits of financial imagination"—that is, the extent to which financial operations are controlled by the ideal of a free investor operating in an efficient market—as a crucial object in the examination of the meaning of considering and treating things as assets, an anthropological path further developed in Muniesa et al. (2017).

Heterodox Political Economy and Economic Geography

Further attempts at tackling the problematics of assetization through a conceptual, economic critique of the notion of financial capital include the idea that contemporary global finance is governed by a specific logic of "fictitious capital" (Durand 2017). Such developments certainly require reconsidering the categories of the Marxian critique of value, as pertinently explored by Ascher (2016). Examples of literature worth considering here include critical and heterodox economics and political economy, as well as economic geography.

Working within a Veblenian tradition, scholars like Herman M. Schwartz, Ronen Palan, and Nitzan and Bichler have all contributed significantly to the dissection of assets and assetization in recent times, even if they do not use these specific terms. In their grand challenge to both neoclassical and Marxist economics, Nitzan and Bichler (2009) aim to rethink capitalism as a mode of accumulation based on capitalization. Drawing on classical and neoclassical ideas about the valuation of capital—such as discounting developed by Irving Fisher—but also on the conceptual idiom of Cornelius Castoriadis, Nitzan and Bichler argue that an array of human activities can be and are capitalized, thereby constituting asset values as an assessment of future earnings, suitably adjusted for risk. As Bjerg (2014) notes, this identification of risk in valuation is a relatively recent phenomenon. However, at this point Nitzan and Bichler argue that while tangible assets can be priced relatively easily, the opposite is true of intangible assets, even though the latter have become increasingly important over time (Palan 2013; Haskel and Westlake 2018). Understanding intangible assets has ended up as a major focus of study for many heterodox economists and political economists as a result, although we can only mention a few here. Perhaps the clearest examples are Schwartz (2016, 2017) and Bryan et al. (2017). In his work, Schwartz is interested in how firms externalize their tangible assets and outsource production as part of a strategy to rely on IPRs as a source of monopoly rents. In their work, Bryan et al. are more interested in the "new ways of *managing and deploying* intangible assets" as part of a broader concern with offshore financial flows, tax evasion, and wealth management (59). It is obvious that both concerns are valid and pressing issues, and it raises the important geographical question regarding assets—namely, how place specific and dependent are they?

Several economic geographers, broadly speaking, have focused on these aspects of assetization, especially as it relates to infrastructure finance and financing (Leyshon and Thrift 2007), although others are now addressing things like carbon finance (e.g., Bridge et al. 2019). An early and notable example of the former is a book by Graham and Marvin (2001) called *Splintering Urbanism* in which they highlighted the increasing fragmentation of urban infrastructure resulting from the privatization of public assets. This unbundling of specific infrastructure assets (such as roads) from an urban system resulted from new forms of asset management and its monetization (Birch and Siemiatycki 2016). More recently, geographers interested in financial geographies have highlighted the reconfiguration of place-based things like infrastructure (e.g., railways) and farmland as a new asset class (e.g., Adisson 2015; O'Brien and Pike 2015; Larder et al. 2017; Ward and Swyngedouw 2018; Ouma 2019), which have become popular with pension funds in particular (Orr 2007), as well as so-called idle assets (e.g., cars, spare rooms, labor) that are brought into economic circulation through new social media platforms (e.g., Uber, Airbnb, TaskRabbit)—see Langley and Leyshon (2017) for example. While some of these arguments are premised on the idea that these sorts of assets are a form of financial capital (e.g., O'Neill 2013), it is noticeable that much of the work highlights the spatial and physical dimensions of assets that the emphasis on finance often obscures. For example, an infrastructure asset can be defined by the immediate costs of its physical construction as much as by the life-cycle costs of its use and maintenance (Birch and Simeiatycki 2016). As such, the technical calculation and accounting of costs and benefits constitute an asset as much as its form, which brings us to accounting.

Critical Accounting Studies and the Sociology of Capitalism

One promising attempt at situating the asset form within a wider assessment of the evolution of capitalism is to be found in recent work by Luc Boltanski and Arnaud Esquerre (2016). The asset appears in their analysis as a paradigmatic form of arranging and displaying objects of valuation, a form that derives analytically from the "collection form" (their main topic of investigation). While the latter enriches the value of things (i.e., it justifies higher prices) by situating them in a context that augments their memorial force (signature, heritage, singularity), the asset form appears in speculative appraisal to be the key to such process of enrichment. Contrary to concomitant perspectives that see in financial valuation a paradigmatic critique of the market (Muniesa et al. 2017), Boltanski and Esquerre base their analysis on a generalization of the commercial transaction as the basic template of the capitalist arrangement.

The specificity of the accounting apparatuses that govern a regime of assetization has been submitted to closer scrutiny by Chiapello (2015). The very notion of financialization, which is often used in reference to securitization, is best understood (according to Chiapello) as an all-encompassing process of "colonization" by finance of various kinds of valuation practices. The asset form is considered here as a properly capitalistic form—that is, a form that presents assetized things as capital rather than as commodities. The connection of this approach to critical perspectives in accounting is obvious; it links to a productive tradition in accounting studies that has examined the genealogies of financial accounting and its uses (see, for example, Miller 1991; Toms 2010; Bryer 2012, 2013a, 2013b).

Science and Technology Studies

Another field that is starting to engage with assets, especially as techno-economic objects, is science and technology studies (STS). As a discipline focused on the coproduction of technoscience and society, STS is been well placed to provide critical, constructivist accounts of assetization, especially in terms of *how* things are turned into assets. Although it is a topic that stretches back several decades (e.g., Levidow and Young 1981), the last few years have witnessed a surge in STS scholars interested in assets as a distinct analytical category relevant to understanding contemporary technoscientific capitalism. A few examples of this work, this book aside, include discussions of intangible assets such as intellectual property (Birch and Tyfield

2013; Lezaun and Montgomery 2015; Martin 2015), which is probably the most prevalent type of analysis, as well as human capital (Cooper and Waldby 2014), business models and valuation practices (Birch 2017a; Muniesa et al. 2017; Doganova 2018a), and new kinds of assets such as personal or health data (Vezyridis and Timmons 2017; Geiger and Gross 2019; Sadowski 2019). While their substantive topics may be highly differentiated, these scholars share a common interest in examining assets as distinct from commodities and as constituted by and coming to constitute, in a performative fashion, value and valuation in technoscientific capitalism.

Notably, these STS scholars seek to engage with the issue of understanding how future value is constructed in the present through techno-economic processes, practices, and knowledge (Doganova 2018a). While there has been growing interest in "promissory economies" in STS (e.g., Petersen and Krisjansen 2015), much of this work tends to assume that it is the inherent, embodied, or latent material qualities of life released by technoscience that generates value. However, as Paul Martin (2015, 425) notes, expectations—and thus value and valuation—are not "spontaneously created…but have to be socially organised through the authoring of expectations and the enrolment of actors." In this sense, assets are made; they do not simply emerge from the ether or the earth. Returning to Chiapello's (2015) work, assetization necessitates the definition of what an asset's boundaries are, the measurement of its quantity and quality, and a valuation of its monetary worth—all of which requires an enormous amount of work by technical experts, economists, valuation analysts, policy-makers, and others.

Paradoxes, Problems, and Open Questions

Assetization studies certainly do not constitute a unified field or a clear-cut research program. They are rather characterized by the conjunction of gray zones and unsettled issues. The aim of this book is to make these explicit and submit the subject matter to collective scrutiny, especially as this relates to the various implications of the asset condition (Muniesa et al. 2017; Birch 2018). Among the unsolved paradoxes and problems that assetization raises, the following seem to be particularly pressing. One is the difference between assetization and commodification and the concomitant problem of the place of markets versus finance in the determination of capitalism.

Another one is the extent to which the dominance of an investment viewpoint in assetization processes runs counter to an entrepreneurship perspective—or alongside. Whether the asset form is essentially oriented toward private appropriation or, on the contrary, is a form that can be articulated in terms of the public interest also constitutes a pending question. The wobbly status of the immaterial in assetization processes is an equally relevant question. And so is, we believe, the legal configuration of these processes and the extent to which they are marked by a transition to contractual regimes of governance, as opposed to more classic forms of ownership.

Financial Valuation and the Problem of the Market

A research agenda on the breadth and boundaries of the processes that fall under the rubric of "assetization" is evidently linked to a wider, critical appraisal of the tendencies of contemporary capitalism. Finance comes here at the forefront, as contemporary capitalism is insistently examined in critical accounts in the light of the conditions and consequences of "financialization." But the critique of capital is also often associated with a critique of markets, with commodification—the commodification of everything—presented as the prime engine of the advancement of capitalism. And still, as the literature presented above tends to suggest, there is something specifically central about market valuation in the type of appraisal that assetization offers. The standard distinction that is done in financial analysis between the fundamental value of an asset (determined with the help of discount methodologies, such as DCF, that place future revenues or earning power at the crux of value) and its market price (which basically depends on the present dynamics of strategy and information among buyers and sellers) is essential to the debate (Ortiz 2014). This market price that the finance-savvy analyst can deem overvalued or undervalued in the light of so-called rational valuation methodologies is often termed "speculative" in both financial parlance and technical literature.

The true value of the asset, as the vernaculars of financial valuation have it, is the one that stems from its capacity to create value (Ortiz 2014; Muniesa 2017). A well-known paradox links both kinds of value through notions of "efficient market," which stipulates that market prices ought to converge toward fundamental value insofar as the market is supposed to be composed of finance-savvy investors who calculate financial value. Still, in some remarked critiques of finance, it is precisely the speculative aspect of market appraisal that is signaled as the prime characteristic of finance (e.g., Orléan 2014; Ascher 2016; Boltanski and Esquerre 2016; Adkins 2018). For example, Konings (2018) has problematized the foundationalist understanding of value that critical conceptions of speculative finance need to rely on (see also Cooper and Konings 2015). The debate remains open on whether—or to what extent—assetization partakes of a logic of commodification and therefore of speculative potential or, rather, participates in a capitalistic logic that aims in part at canceling the threat that a market logic poses to the security of revenue appropriation.

Debates around notions of biocapital and bioeconomy are also pivotal in the articulation of this question. Once considered the main angle of the problem, the commodification of biological entities seems to give way to situations of massive value creation in which the commodity form is less important. The key STS scholars writing in this area, including Kaushik Sunder Rajan, Nikolas Rose, Catherine Waldby, and Melinda Cooper, tend to frame discussions around speculative or promissory value, implying that value is largely speculation—a bet by investors, essentially (Birch and Tyfield 2013). Critics of these thinkers point out that some biotech firms, like Amgen, do create value—although how is an important issue to consider—while other biotech firms are primarily financial "artifacts" (Mirowski 2012; Birch 2017a, 2017b). The latter might imply some speculation, but it is premised on an innovation regime in which risk is outsourced (Glabau 2017), necessitating the careful management of value and valuation (Birch 2017a) and the deployment of a business model culture for which assetization is a distinctive objective (Doganova and Muniesa 2015). The very notion of assetization has already been explicitly used by Geiger and Gross (2019) in order to address precisely those issues in the case of the consumer genomics testing industry where firms engage in new forms of health data brokerage.

Thinking like an Investor, Thinking like an Entrepreneur

Locating in the investor's viewpoint the matrix from which assetization makes sense highlights an alternative viewpoint that is increasingly sidelined—namely, that of the entrepreneur. Both personas certainly feature as proxies for complex institutional formations, and both represent central characters in the narratives that dominate the jargon of value creation (Muniesa 2017). In many vernacular (and scholarly) accounts of contemporary business life, the entrepreneur may appear as a virtuous character fighting a moral fight against an investor-qua-rentier that stifles the creation of real value with a short-term, purely financial view (Mazzucato 2018). In other accounts, it is a visionary investor (e.g., the archetypical venture capitalist or "business angel") who stimulates the entrepreneur on the path to economic success. But to what extent do these ideal types oppose each other? To what extent do they form a cultural complex that assetization requires in order to *make sense* as a moral and political narrative? To what extent does this narrative leave other characters out of the value narrative? As assetization becomes the template according to which value creation is gauged, the investor and the entrepreneur-qua-investor become the dominant ingredients of that narrative, with other figures such as the state or publics being left out (or transformed accordingly).

Assetization is, as much as anything else, a process of narrative transformation. Things are accounted in terms of the asset, but social actors participating in economic activity are also reframed, altered, and conceptualized anew. Understanding these shifting visions and identities is critical, since it is the techno-economic practices of social actors that define an asset's boundaries, measure its characteristics, and evaluate its worth. An especially important transformation, in this regard, is the take-up of financial logics by social actors, including managers and executives, government officials, individual consumers, and more. Chiapello (2015), for example, has highlighted how mainstream valuation approaches in modern finance all share the same goal of taking the investor viewpoint as the basis for assessing the societal value of any investment based on the assumption that "investors—thanks to their special knowledge—are seen as the most capable of allocating available economic resources in an optimal way" (19). This idea, more fully developed by Ortiz (2014; see also Muniesa et al. 2017), can be also observed beyond the technical realm of financial valuation. It is, for instance, the investor that stands as the prime semiotic engine around which revolve the efficacy of business models (Doganova and Muniesa 2015).

In an article titled "Seeing like a Market," Fourcade and Healy (2017, 10) argue that the "digital economy's classificatory architecture allows market institutions to apprehend their clients, customers, or employees through new instruments of knowledge, efficiency and value extraction." As Fourcade and Healy frame it, markets increasingly force individuals to accrue "übercapital," representing a narrative resource—or asset—resulting "from one's position and trajectory according to various scoring, grading and ranking methods" (14). Whether übercapital is more of an asset for individuals themselves or for the companies collecting the data that constitutes übercapital is hard to disentangle—it seems to be both, according to the authors—which offers us an insight into the complexities underlying assetization. This concept highlights how different entities—imaginary or not—are enrolled in the pursuit of specific narrative and objectives (see also Sadowski 2019).

Managers and executives are exhorted to "think like an investor" rather than as entrepreneurs; as a result, they bind themselves to thinking and acting like an investor (such as adopting "opportunity cost" thinking). A recent *Vital Signs* report from Ernst & Young (2015, 5) stresses the need for company executives "to think more like investors than managers." Paradoxically, this implies that in order to avoid "activist" shareholder interventions they do not want, executives have to think more like shareholders in the first place. It goes without saying that thinking like an investor entails equating value creation with return on investment rather than with the research or development of new products and services (Birch 2017d; Muniesa 2017; Glabau 2017). Assetization is linked here with the prevalent power of the investor, raising the question of whether there is an alternative to seeing the world like an investor. The very notion of the "investee" displacing the "worker" as the main political identity in matters of resistance against financialization, as examined by Michel Feher (2018), is certainly part of this syndrome. Attempts at capturing (and reclaiming) the role of the state in value processes are also pivotal in the conversation. Proposals for the study and articulation of notions of state-driven entrepreneurialism are promising in this respect (see, for example, Mazzucato 2013; Goldstein and Tyfield 2018). But how far can they go in questioning the investor regime that seems to control the spread of assetization?

Private Profit and the Public Interest

A social inquiry into assetization certainly meets in the private interest a crucial driver. Considering something in terms of an asset usually means preparing it for the appropriation by an investor (shareholder or creditor) of the revenue—the rent—that derives from the entitlement (stock, bonds, or other investment forms). But still, many of the traits that we recognize in processes of assetization can be recognized too in contemporary developments in public finance. The rationalization or modernization of public investment and public management—whose aim remains the realization (or rather optimization) of the public interest—can rely, sometimes abundantly, on the methodologies that financial valuation offers. The diffusion in advanced liberal democracies of policy doctrines and operational tools derived from public choice paradigms or from management accounting is certainly a topic that an inquiry into assetization might capture.

As competitiveness becomes the driving rationale of neoliberal public policy (Davies 2014), the urge to attract investors becomes a crucial political imperative. The evaluation of large infrastructural projects that take into account intergenerational responsibility or long-term effects is more and more dependent on discount methods that require thinking of expenditures as an asset form—for example, in the case of nuclear waste management (Muniesa et al. 2017; Saraç-Lesavre 2017, 2018). The extent to which the neoliberal development often dubbed "New Public Management" is an extension of the asset condition (taxpayer as investor, public budget as investment fund, public holdings as capital assets, public service as return) is indeed a central part of the agenda for assetization studies. Analyzing how the nature, scope, and function of the state are made explicit in different manners through changing performance indicators furthers the understanding of the way in which the allure of the asset affects the political meaning of the state (Muniesa and Linhardt 2011; Mennicken and Lodge 2015; Mennicken and Muniesa 2017).

Multiple research objects stand as obvious sites in which this problem can be explored in depth. One area of prime interest to an STS agenda is the governance of science. The connections that are established in government bodies and state agencies between public policies on scientific research and economic objectives (e.g., technological innovation, economic growth, and value creation) are marked by the diffusion of metrics and models that tend to describe scientific value in terms of a future return (Robson 1993, 1994; Godin 2004, 2007; Muniesa and Linhardt 2011). The fact that public funding bodies look at innovation projects through the lens of the business model is also of particular relevance (Doganova and Eyquem-Renault 2009). Other topics such as the development of entanglements with private-sector practices in order to ensure the economic viability of public service are equally relevant. Public-private partnerships (PPP) are a particular interesting case, as their failures or successes need to be made sense of, besides their effective or ineffective delivery, from the perspective of the alteration of the balances of strength between states and private actors (Birch and Siemiatycki 2016). The rapid development of social impact bonds and other forms of impact investment in the modernization of public social policies also provides a fertile ground on which to explore how the object of public service (e.g., the homeless) can be turned into an asset (Cooper et al. 2016). Contemporary practices of "philanthrocapitalism," which in part also rely on the idea that an investor's viewpoint is best positioned to see where money can do most good, add to this accumulation of situations in which the assetization of public policy can be observed (McGoey 2015). The key question of what it means to interpret these shifts and practices in terms of a crisis of political sovereignty is still open to scrutiny and debate.

The Material and the Immaterial

Notions of incorporeal or intangible property were already crucial in the theorization of the asset form by the likes of Veblen (1908a, 1908b) and Commons (1924). And this discussion stretches back into history. In distinguishing between incorporeal (i.e., financial) and intangible (e.g., goodwill, brands, reputation) assets, Ronen Palan (2013) points out that concepts of "intangible property"—to use his term—can be found as far back as late sixteenth-century English court rulings, especially as it related to trust and trustworthiness in a person or their products. As for today, see Helms et al. (2011) on "reputation management." Assets come in all sorts of shape, size, tangibility, and corporeality, potentially confusing our discussion even further. And the distinction between materiality and immateriality seems to govern, to quite an extent, the conversation on what assets are or should be about. The extent to which these distinctions rely on prephilosophical or nonconstructivist ideas of materiality remains debatable (Muniesa 2016). But the physical materiality or immateriality of an asset is certainly—and interestingly—used in the literature in a manner that highlights a series of questions worth considering. Intangible assets are also increasingly highlighted as the primary source of company value (Nitzan and Bichler 2009; Birch 2015; Bryan et al. 2017; Haskel and Westlake 2018), prompting us to ask how we might differentiate between asset forms and assetization on the basis of their materialities. To use simple binaries, an asset can be large/small, complex/simple, heterogeneous/homogenous, perpetual/exhaustible, rivalrous/nonrivalrous, and so on.

While the physical materialities theorized by economic geographers (e.g., Adisson 2015; O'Brien and Pike 2015; Christophers 2016; Ouma 2019) might seem relatively clear-cut—for example, a railway is a physical piece of equipment that cuts across national and international landscapes and borders—this belies the complexity at play in these examples. As Birch and Siemiatycki (2016) note, certain socio-technical structures and systems are easier to unbundle and assetize than others; for example, a road with easily tolled entry and exit points is easier to monetize than a regional railway system with multiple and multilayered entry and exit points. Similar concerns could be raised regarding any other tangible asset: What are the physical constraints on monetization? Or capitalization? Each is potentially unique, as a result of its siting and spatiality, meaning that turning these things into assets also entails a huge supporting cast of social actors, including accountants, engineers, analysts, financiers, and so on, able to make valuations on a social basis (Chiapello 2015). All of this, it is worth emphasizing, is complicated further when looking at intangible assets.

As we and the contributors to this book are arguing, almost anything can be turned into an asset given the right techno-economic configuration. Even personality sells; for example, the YouTube star PewDiePie has more than 100 million subscribers and earns millions of dollars every year by (without wanting to denigrate the effort he puts into it) playing computer games and making gurning noises. A growing chorus of critical political economists, particularly drawing on ideas in autonomist Marxism, have sought to theorize the growing importance of intangible assets in a range of emerging capitalist forms and formations constituted by immaterial, cognitive, and affective labor (Moulier Boutang 2011; Marazzi 2011). Building on broader societal and policy discourses around the shift to a knowledge economy (Luque 2001), these thinkers focus on the extension of ownership and control over knowledge, emotions, and socialities as part of a process which Moulier Boutang (2011, 14) describes as the "systematic conversion of rent positions in intellectual activities into tradable assets." It is not all theoretical either, with international agencies like the OECD promoting the idea of "knowledge markets," "knowledge-based capital," and "intellectual property assets" (OECD 2008, 2010, 2012). Likely inspired by the expansion of the Internet, there has been a major shift in the theoretical and social understandings of knowledge—including research, creative works, and user-generated content—as both private property *and* public commons

(Arvidsson and Colleoni 2012). This has led to ongoing conflict and contestation between multiple social actors, such as hackers, hobbyist groups, patent offices, trade organizations, multinational corporations, and governments (Hope 2008). As such, examining how things are turned into assets means understanding how material and immaterial assets are maintained or challenged as such.

Ownership and Property versus Contracts and Licenses

The fact that turning something into an asset is in part a juristic operation is certainly well established within a tradition of legal scholarship that examines the constitutive role of law in capitalism (Deakin et al. 2017; Pistor 2019). Doctrines of investor protection, as developed in the tradition of the economic or financial analysis of the law, are an important element in the establishment of the juristic contexts in which the asset condition can thrive (La Porta et al. 1988; but see Pistor 2013). But assetization is increasingly framed by a shift from regimes of ownership to regimes of contract. Ownership rights are often described as a bundle of rights, including exclusion, use, sale, and so on (Hodgson 2003; Kang 2015). Despite their potential complexity, what marks them out is that they are publicly constituted by state legislative action and legal enforcement; for example, it is not possible to limit the future use of property after it has been transferred (Perzanowski and Schultz 2016). Focusing on real or personal property, many of these rights are specific to tangible things like land, housing, apparel, cars, and so on.

More recently, the expansion of intangible assets has engendered a transformation in intellectual property rights (IPRs) to protect those assets. Such IPRs cover things like copyright, patents, and trademarks, and represent the main way that companies have sought to ensure that they can avoid the effects of the marginal zero cost of production (Rifkin 2014)—namely, once an initial investment in production has happened, subsequent costs of production tend toward zero. As content, in whatever form, has become almost costless to reproduce, companies and individuals have turned to IPRs to secure their profits (Schwartz 2017). According to Perzanowski and Schultz (2016), the ongoing institutional transformation of ownership in intangible assets has led to a shift in juristic operations from property rights to contract law, represented specifically by license agreements.

License agreements, while remaining obscured behind a rhetoric of ownership, reflect a transformation in the ownership and control of "purchased" goods and services. Really, consumers are simply licensing things (such as a downloaded music track) produced with intellectual property (such as a copyright). As Birch (2016) and Perzanowski and Schultz (2016) argue, however, the value of these intangible assets is constituted by the contractual costs they impose on the public, including access limits, licensing complexity, and anticompetitive effects. For example, it is difficult for consumers to switch technologies (e.g., Apple) after they have spent so much on complementary products and services (e.g., iTunes music). As a result, the rights of asset holders come to trump consumers, who can end up locked into one asset enclave economy or another, or platform owners extract rent from other people's assets rather than their own (e.g., Uber) (Birch 2020). Such consequences are some of the reasons that many people turn to open source (or open science) mechanisms as a way to coordinate the social organization of research and development (Benkler 2002).

Turning Things into Assets

The collection of studies gathered in this book originated in an open call for proposals for a panel on "Turning Things into Assets" at the 4S/EASST 2016 conference in Barcelona. The Society for Social Studies of Science (4S) and the European Association for the Study of Science and Technology (EASST) are two major scholarly associations established in the academic field of STS, but their conferences also attract researchers well beyond that field. The "Turning Things into Assets" panel constituted a space in which the interdisciplinary dialogue we referred to above took place. And the contributions that are included in this book provide, we believe, palpable evidence of the emergence of a fruitful conversation. The chapters are organized into four sections that roughly correspond to four broad and interrelated topics on how different things are turned into assets: knowledge, infrastructure, nature, and publics.

Turning Knowledge into Assets

The OECD's lingo on the "valuation and exploitation" of intellectual property constitutes a case in point for understanding the operations that the assetization of knowledge requires, as Hyo Yoon Kang suggests in her contribution. This is entirely about refining the contours of the intangible asset (e.g., Kamiyama et al. 2006), and these contours are essentially juristic. The analysis Kang offers can be read as a theory of the *abstract tangibility* that the law enables for intangible knowledge. Patents configure knowledge with the material, objectified contours that the asset condition requires. That this comforts the now-dominant view that "wealth mainly lives in intellectual property," to quote the *Financial Times* comment cited by Kang (Foroohar 2017), is clear. The knowledge economy stands certainly as the medium in which the value creation function of intellectual property is most visible. The emergence of patent portfolios as a legal problem (Risch 2013), Kang tells us, illustrates the extent to which the investment logic of patents as financial assets introduces novel, forward-looking concerns that can be disconnected from the logic of the actual use or actual commercial potentiality of the patented invention.

The rise of "platform capitalism," to use the expression developed by Srnicek (2017) and Langley and Leyshon (2017), certainly comes into the picture too when we try to grasp the business model culture of the data-driven economy (see Sadowski 2019). Data sets—or "datassets" as Thomas Beauvisage and Kevin Mellet call them in their contribution—here stand as the paramount asset form. The talk of an emerging "asset class" in relation to personal data, Beauvisage and Mellet claim, serves as an indication of the political concern with the "untapped opportunities for socioeconomic growth" and "the importance of collecting, aggregating, analysing and monetising personal data" (World Economic Forum 2011, 7). Beauvisage and Mellet focus on the success and vicissitudes of behavioral data capture: data brokers, tracking technologies ("cookies"), and data management platforms. They provide assetization studies with a useful connection to marketing studies, especially of the kinds that have focused on the "cultural economies" of market attachment (McFall 2014; Cochoy, Deville, and McFall 2017). They observe, though, a fundamental contrast between commodification and assetization: monetizing personal data may mean at some point selling these personal data away, but as the relative failure of consumer-to-business schemes in personal data illustrates, that strategy tends to be replaced by an investment, forward-looking, rent-seeking rationale.

Emphasis on the future (expectations in the form of future value) is at the center of the asset condition (Doganova 2018a). This is an insight that Victor Roy develops in his contribution on the political economy of biomedical innovation. The logic of the "pharmaceutical asset," he claims, increasingly controls the dynamics within the industry. The spread and articulation of the doctrine of shareholder value maximization create, according to Roy, quite a specific situation. Rather than being evaluated in the light of their current profitability, large pharmaceutical companies are considered through their potential to deliver future earnings for shareholders. This translates into a somewhat paradoxical form of *limitation*—if not cancellation—of the very finality of pharmaceutical innovation, which is (or ought to be) to develop cures *in the present*. The way in which this movement affects biomedical knowledge and biomedical reality at large (starting with the patient's) is patent in Roy's investigation. Current sociological research on the nature of economic expectations and on the spread of future narratives finds here a crucial test bed for the examination of the *futures* that assetization both produces and requires (Beckert 2016; Beckert and Bronk 2018).

Turning Infrastructures into Assets

That infrastructures are mundanely relational is today commonsense in the STS-inspired social sciences (Star 1999). But what about the financial investment that infrastructures are the product of? Its *relational nature* needs to be captured too. If an infrastructure is a compound of localized, hybrid elements, all the more so when that infrastructure adopts the asset form. In their contribution, Alain Nadaï and Béatrice Cointe draw from the Callonian vocabulary of the "agencement" (Callon 2016). Examining the case of environmentally motivated co-ops investing in mutualized photovoltaics or wind farms, they observe the multiple territorial and spatial dimensions of the future value potential of the investment projects. Of particular importance is the role the state plays in such assemblages: feed-in tariffs ensure the stability and predictability of the revenue stream that make the entire project economically viable. This role is not without relevance to the *safe ground* that the state ought to provide, in general, to the investor's gaze through the notion of the "risk-free asset" (Boy 2015). And it is also connected to the emergence of a society of engaged, conscientious stakeholders that adopt the shape of a "society of investors" (Davis 2009).

Railroads, an infrastructure crucially dependent on (and interlocked with) the state, offer promising opportunities for furthering the examination of the shapes of infrastructural assetization. Timothy Mitchell has signaled how railroads have played a central role in the material history of the construction of "durable structures of accumulation where a certain amount of the income that can be expected in the future is sold to investors in the present" (Abourahme and Jabary-Salamanca 2016, 740). In her contribution, Natalia Buier further explores that type of angle. A contemporary case study of an ambitious high-speed rail infrastructure program in Spain allows her to expose the essentially unstable aspects of the process of turning railway transportation into a viable, competitive asset. The program was abundantly contested, and the recourse to "respectable" economic metrics such as cost-benefit analysis merely rendered more explicit the political nature of the project—Buier draws here a parallel with Porter (1995). As a process, the assetization of infrastructures requires the constant production of its conditions of possibility. It also involves, rather than a straightforward process of privatization, an ongoing reconfiguration of the public sector.

Turning Nature into Assets

Offering the keys of the valuation and management of natural resources to properly equipped investment managers and financial analysts is a movement that can be identified in several relevant episodes in the history of financial valuation. This is certainly the case of forestry, where the establishment of rotation and yield precipitated remarked-upon innovations, such as DCF methods (Doganova 2018a, 2018b; Muniesa et al. 2017). Mining too; in his contribution, Paul Robert Gilbert ethnographically explores the valuation talk and work that underpins the mineral exploration investment industry. The material side of assetization processes is highlighted, as managing the assets means literally determining where to excavate and how. The sophistication of project finance in this area has also prompted valuation controversies—for example, through the attempted introduction of "real options" analysis—that further stimulates an assetization gaze, as Gilbert shows. But the crux of an assetization analysis can also reside in the examination of the political identifications these techniques accompany. Who are the enemies of value creation in this complex of metrics and narratives? Gilbert emphasizes the rise of resource nationalism as a major threat and the ethnographic record of the menace of the "next Venezuela," in reference to the fragile nationalization of mining business in some Latin American jurisdictions. Political risk indeed becomes one crucial ingredient of the risk rhetoric of assetization (Boy 2015; De Goede 2005) and the call for "durable legal foundations" the juristic cement of the asset condition (Commons 1924).

Natural entities, however, tend to blur a too-sharp analytical distinction between the commodity and the asset forms. Once it enters the realm of farming, a cereal seed, for example, can obviously work as a commodity good that can be bought and sold at a price in the market but also as the source of a future yield (i.e., asset). Veit Braun explores this shifting condition in his contribution, using the detaching properties of market transactions as a compass. One well-known formal property of market goods is their alienation capacity—that is, the propensity to make the parties to the transaction "quits" (Polanyi 1944), a propensity often compared in economic anthropology to the bonds, obligations, and attachments prompted by exchanged gifts (Carrier 1995; Godelier 1999). The asset, though, is certainly characterized, too, as the gift (anthropologically understood), by its capacity to maintain attachments between investor and investee so as to secure a future yield, rent, or return. Braun rightly confirms the importance that patent law has in the determination of the conditions in which seeds ought to be protected as assets. Patent politics, assetization studies ought to definitely note, is a crucial ingredient of the way in which natural resources and life-forms are made economically, capitalistically accountable (Parthasarathy 2017).

Turning farmland into an investable thing has gained relevance as a global prime concern in recent years, calling for a series of technoscientific (but also legal and political) measures that have not gone unnoticed in assetization studies (Ducastel and Anseeuw 2017; Fairbairn 2014; Li 2014, 2017; Larder et al. 2018; Ouma 2019). A more overarching idea is also gaining momentum in the face of the rush to develop economic solutions to the environmental crisis—namely, that of considering all environmental resources, or nature altogether, as an asset or pool of assets. In his contribution, Les Levidow deconstructs the interplay of capitalistic metaphors that accompany the establishment of novel (and much touted) templates such as natural capital accounting (see also Akerman 2005; Moore 2015; Coffey 2016; Muniesa et al. 2017). Emphasis on the power of the metaphor of nature as capital enables an understanding of the cultural dimension of assetization processes. But as the communication materials put forward at the World Economic Forum on Natural Capital clearly demonstrate, there is an essential political dimension to this, since emphasis on the asset means emphasis on the expertise and legitimacy of the asset manager in determining which parts of nature should be valued or not (see Felli 2014). Levidow signals in particular how this process can disorient critical voices that see in the commodification of nature the prominent menace of capitalism and that (inadvertently or not) help promote a revaluation of nature (meaning a revaluation as asset) in order to protect it from the perils of market dilapidation. The convergence in language that the capital metaphor prompts might serve well as an environmentalist strategy, but it also precipitates, Levidow shows, a depoliticizing focus on reputational concerns.

Turning Publics into Assets

The rise of an "asset base" rhetoric in the provision of social services reflects another key object of inquiry for assetization studies (Langley 2006, 2008). The UK often features as an example of this trend, probably due to the intimate experience of this jurisdiction with neoliberal policies (Davies 2014; Springer et al. 2016; Birch 2017c). And British higher education is certainly well ranked among the topics suitable for this elucidation (Mennicken and Muniesa 2017). The introduction of performance measurements (such as the Research Excellence Framework and the Teaching Excellence Framework), variable tuition fees, and autonomy in university governance have led to the development of a British "new cultural epoch of managerialism" (Shore and Wright 2000). The extent to which this is a cultural epoch of assetization requires that we examine how higher education and its publics are transformed (or not) into the asset form. Sveta Milyaeva and Daniel Neyland answer this question in their contribution. Their investigation of the introduction and consequences of income-contingent repayment loans for university students in England and Wales illustrate the way in which such reforms involve important shifts in the accounting view, with some things (i.e., loans) not being categorized as spending anymore but finding their way into the asset column of the accounting imagination. They also observe how the assumption made about the discount rate becomes a key driver of the policy debate (Britton and Crawford 2015). Alongside the naturalization of net present value, the politics of discounting definitely strikes us as a defining feature of the political trajectory of the asset form and condition in public services.

Technoscientific capitalism entails a wide variety of financial instruments that rely on the identification of "monetizable social ills," to cite an informant quoted by James Williams in his contribution, as a way to improve social services. Williams scrutinizes the case of the investment-based funding model known as social impact bonds (SIBs). These recent financial instruments already stand as an intriguing research object for studies of financialization and marketization (see Cooper et al. 2016; Neyland 2018). However, Williams argues that the realities of the SIB industry are difficult to square with the financialization and marketization narratives featured in many critical accounts. A large study, which includes interviews with SIB developers and investors in Canada, the US, and the UK, enables Williams to locate the core of an assetization approach to the subject matter in the problem of valuation. The emergence of a genuine concern for "evaluation risk" (namely, the influence that particular benchmarking and impact assessment procedures such as randomized controlled trials can have on the very profitability of the investment scheme) provides a vivid illustration of the centrality of that problem. How can impact be properly defined and deadweight properly controlled, for whom and by whom, ask SIB practitioners? The fact that return on investment is typically dependent on social phenomena (recidivism, rough sleeping, unemployment) that need to translate into measurable individual behavior also constitutes a unique valuation challenge. Beyond discussions of whether a SIB can be considered a viable asset (a question that raises the issue of its liquidity), we perceive how the public's "social ills," which definitely need to acquire the shape of a delineated human behavior, adopt the contours of a monetizable asset form—or, in other words, of an investee condition (Feher 2018).

References

Abourahme, N., and Jabary-Salamanca, O. 2016. Thinking against the Sovereignty of the Concept: A Conversation with Timothy Mitchell. *City* 20 (5): 737–754.

Adisson, F. 2015. The Assetization of State Places: Insights from the Railway Properties in France and Italy. Paper presented at AAG Annual Meeting, Chicago, IL.

Adkins, L. 2018. *The Time of Money*. Stanford, CA: Stanford University Press.

Adkins, L., Cooper, M., and Konings, M. 2019. Class in the 21st Century: Asset Inflation and the New Logic of Inequality. *Environment and Planning A*, doi:10.1177/0308518X19873673.

Aitken, R. 2007. *Performing Capital: Toward a Cultural Economy of Popular and Global Finance*. New York: Palgrave Macmillan.

Akerman, M. 2005. What Does "Natural Capital" Do? The Role of Metaphor in Economic Understanding of the Environment. *Environmental Education Research* 11 (1): 37–52.

Arvidsson, A., and Colleoni, E. 2012. Value in Informational Capitalism and on the Internet. *Information Society: An International Journal* 28 (3): 135–150.

Ascher, I. 2016. *Portfolio Society: On the Capitalist Mode of Prediction*. New York: Zone Books.

Baars, G., and Spicer, A., eds. 2017. *The Corporation: A Critical, Multi-Disciplinary Handbook*. Cambridge: Cambridge University Press.

Beckert, J. 2016. *Imagined Futures: Fictional Expectations and Capitalist Dynamics*. Cambridge, MA: Harvard University Press.

Beckert, J., and Bronk, R., eds. 2018. *Uncertain Futures: Imaginaries, Narratives and Calculation in the Economy*. Oxford: Oxford University Press.

Benkler, Y. 2002. Coase's Penguin, or, Linux and *The Nature of the Firm*. *Yale Law Review* 112: 369–446.

Berman, E. P. 2012. *Creating the Market University How Academic Science Became an Economic Engine*. Princeton, NJ: Princeton University Press.

Beunza, D., and Stark, D. 2004. Tools of the Trade: The Socio-technology of Arbitrage in a Wall Street Trading Room. *Industrial and Corporate Change* 13 (2): 369–400.

Birch, K. 2015. *We Have Never Been Neoliberal: A manifesto for a Doomed Youth*. Winchester: Zero Books.

Birch, K. 2016. Market vs. Contract? The Implications of Contractual Theories of Corporate Governance to the Analysis of Neoliberalism. *Ephemera: Theory & Politics in Organization* 16 (1): 107–133.

Birch, K. 2017a. Rethinking *Value* in the Bio-economy: Finance, Assetization and the Management of Value. *Science, Technology and Human Values* 42 (3): 460–490.

Birch, K. 2017b. Financing Technoscience: Finance, Assetization and Rentiership. In *The Routledge Handbook of the Political Economy of Science*, edited by D. Tyfield, R. Lave, S. Randalls, and C. Thorpe, 169–181. London: Routledge.

Birch, K. 2017c. *A Research Agenda for Neoliberalism*. Cheltenham: Edward Elgar.

Birch, K. 2017d. Techno-economic Assumptions. *Science as Culture* 26 (4): 433–444.

Birch, K. 2018. What Is the Asset Condition? *European Journal of Sociology* 59 (3): 500–506.

Birch, K. 2020. Technoscience Rent: Toward a Theory of *Rentiership* for Technoscientific Capitalism. *Science, Technology and Human Values* 45 (1): 3–33.

Birch, K., and Siemiatycki, M. 2016. Neoliberalism and the Geographies of Marketization: The Entangling of State and Markets. *Progress in Human Geography* 40 (2): 177–198.

Birch, K., and Tyfield, D. 2013. Theorizing the Bioeconomy: Biovalue, Biocapital, Bioeconomics or…What? *Science, Technology and Human Values* 38 (3): 299–327.

Bjerg, O. 2014. *Making Money: The Philosophy of Crisis Capitalism.* London: Verso.

Boltanski, L., and Esquerre, A. 2016. The Economic Life of Things. *New Left Review* 98: 31–54.

Boy, N. 2015. Sovereign Safety. *Security Dialogue* 46 (6): 530–547.

Bridge, G., Bulkeley, H., Langley, P., and van Veelen, B. 2019. Pluralizing and Problematizing Carbon Finance. *Progress in Human Geography*, doi.org/10.1177/0309132519856260.

Britton, J., and Crawford, C. 2015. Estimating the Cost to Government of Providing Undergraduate and Postgraduate Education. *IFS Reports R105.* London: Institute for Fiscal Studies.

Bryan, D., Raffert, M., and Wigan, D. 2017. Capital Unchained: Finance, Intangible Assets and the Double Life of Capital in the Offshore World. *Review of International Political Economy* 24 (1): 56–86.

Bryer, R. 2012. Americanism and Financial Accounting Theory—Part 1: Was America Born Capitalist? *Critical Perspectives on Accounting* 23 (7-8): 511–555.

Bryer, R. 2013a. Americanism and Financial Accounting Theory—Part 2: The "Modern Business Enterprise," America's Transition to Capitalism, and the Genesis of Management Accounting. *Critical Perspectives on Accounting* 24 (4-5): 273–318.

Bryer, R. 2013b. Americanism and Financial Accounting Theory—Part 3: Adam Smith, the Rise and Fall of Socialism, and Irving Fisher's Theory of Accounting. *Critical Perspectives on Accounting* 24 (7-8): 572–615.

Burton, G. F., and Jermakowicz, E. K. 2015. *International Financial Reporting Standards: A Framework-Based Perspective.* London: Routledge.

Callon, M. 2016. Revisiting Marketization: From Interface-Markets to Market-Agencements. *Consumption Markets & Culture* 16 (1): 17–37.

Carrier, J. G. 1995. *Gifts and Commodities: Exchange and Western Capitalism since 1700.* London: Routledge.

Chiapello, E. 2007. Accounting and the Birth of the Notion of Capitalism. *Critical Perspectives on Accounting* 18 (3): 263–296.

Chiapello, E. 2015. Financialisation of Valuation. *Human Studies* 38 (1): 13–35.

Christophers, B. 2015. The Limits to Financialization. *Dialogues in Human Geography* 5 (2): 183–200.

Christophers, B. 2016. For Real: Land as Capital and Commodity. *Transactions of the Institute of British Geographers* 41 (2): 134–148.

Coase, R. 1937. The Nature of the Firm. *Economica* 4 (16): 386–405.

Cochoy, F., Deville, J., and McFall, L., eds. 2017. *Markets and the Arts of Attachment*. London: Routledge.

Coffey, B. 2016. Unpacking the Politics of Natural Capital and Economic Metaphors in Environmental Policy Discourse. *Environmental Politics* 25 (2): 203–222.

Commons, J. R. 1924. *Legal Foundations of Capitalism*. New York: Macmillan.

Cooper, C., Graham, C., and Himick, D. 2016. Social Impact Bonds: The Securitization of the Homeless. *Accounting, Organizations and Society* 55: 63–82.

Cooper, M. 2017. *Family Values: Between Neoliberalism and the New Social Conservatism*. New York: Zone Books.

Cooper, M., and Konings, M. 2015. Contingency and Foundation: Rethinking Money, Debt, and Finance After the Crisis. *South Atlantic Quarterly* 114 (2): 239–250.

Cooper, M., and Waldby, C. 2014. *Clinical Labor: Tissue Donors and Research Subjects in the Global Bioeconomy*. Durham, NC: Duke University Press.

Davies, W. 2014. *The Limits of Neoliberalism: Authority, Sovereignty and the Logic of Competition*. London: Sage.

Davis, G. F. 2009. *Managed by the Markets: How Finance Re-shaped America*. Oxford: Oxford University Press.

Deakin, S., Gindis, D., Hodgson, G. M., Huang, K., and Pistor, K. 2017. Legal Institutionalism: Capitalism and the Constitutive Role of Law. *Journal of Comparative Economics* 45 (1): 188–200.

De Goede, M. 2005. *Virtue, Fortune, and Faith: A Genealogy of Finance*. Minneapolis: University of Minnesota Press.

Deville, J. 2015. *Lived Economies of Default: Consumer Credit, Debt Collection and the Capture of Affect*. London: Routledge.

Dewing, A. S. 1920. *The Financial Policy of Corporations*. New York: Ronald Press.

Doganova, L. 2018a. Discounting and the Making of the Future: On Uncertainty in Forest Management and Drug Development. In *Uncertain Futures: Imaginaries, Narratives and Calculation in the Economy*, edited by J. Beckert and R. Bronk, 278–297. Oxford: Oxford University Press.

Doganova, L. 2018b. Discounting the Future: A Political Technology. *Economic Sociology* 19 (2): 4–9.

Doganova, L., and Eyquem-Renault, M. 2009. What Do Business Models Do? Innovation Devices in Technology Entrepreneurship. *Research Policy* 38 (10): 1559–1570.

Doganova, L., and Muniesa, M. 2015. Capitalization Devices: Business Models and the Renewal of Markets. In *Making Things Valuable*, edited by M. Kornberger, L.

Justesen, A. Koed Madsen, and J. Mouritsen, 190–215. Oxford: Oxford University Press.

Dreyfuss, R., and Frankel, S. 2015. From Incentive to Commodity to Asset: How International Law Is Reconceptualizing Intellectual Property. *Michigan Journal of International Law* 36 (4): 557–602.

Ducastel, A., and Anseeuw, W. 2017. Agriculture as an Asset Class: Reshaping the South African Farming Sector. *Agriculture and Human Values* 34 (1): 199–209.

Du Gay, P., and Pryke, M., eds. 2002. *Cultural Economy: Cultural Analysis and Commercial Life*. London: Sage.

Durand, C. 2017. *Fictitious Capital: How Finance Is Appropriating Our Future*. London: Verso.

Ernst & Young. 2015. *Vital Signs: How "Fit" Is Your Capital Allocation Strategy?* Boston: EY, LLP.

Fairnbairn, M. 2014. "Like Gold with Yield": Evolving Intersections between Farmland and Finance. *Journal of Peasant Studies* 41 (5): 777–795.

Feher, M. 2018. *Rated Agency: Investee Politics in a Speculative Age*. New York: Zone Books.

Felli, R. 2014. On Climate Rent. *Historical Materialism* 22 (3-4): 251–280.

Fields, D. 2019. Automated Landlord: Digital Technologies and Post-crisis Financial Accumulation. *Environment and Planning A*, doi.org/10.1177/0308518X19846514.

Fisher, I. 1906. *The Nature of Capital and Income*. New York: Macmillan.

Fisher, I. 1907. *The Rate of Interest: Its Nature, Determination and Relation to Economic Phenomena*. New York: Macmillan.

Fourcade, M., and Healy. K. 2017. Seeing like a Market. *Socio-Economic Review* 15 (1): 9–29.

Fourcade, M., and Khurana, R. 2017. The Social Trajectory of a Finance Professor and the Common Sense of Capital. *History of Political Economy* 49 (2): 347–381.

Frase, P. 2016. *Four Futures: Life After Capitalism*. London: Verso.

Fridman, D. 2017. *Freedom from Work: Embracing Financial Self-Help in the United States and Argentina*. Stanford, CA: Stanford University Press.

Foroohar, R. 2017. Big Tech Makes Vast Gains at Our Expense. *Financial Times*, September 17, 2017.

Fuller, S. 2002. *Knowledge Management Foundations*. Boston: Butterworth-Heinemann.

Geiger, S., and Gross, N. 2019. A Tidal Wave of Inevitable Data? Assetization in the Consumer Genomics Testing Industry. *Business & Society*, doi:10.1177/0007650319826307.

Glabau, D. 2017. Conflicting Assumptions: The Meaning of Price in the Pharmaceutical Economy. *Science as Culture* 26 (4): 455–467.

Godelier, M. 1999. *The Enigma of the Gift*. Princeton, NJ: Princeton University Press.

Godin B. 2004. The Obsession for Competitiveness and Its Impact on Statistics: The Construction of High-Technology Indicators. *Research Policy* 33 (8): 1217–1229.

Godin B. 2007. Science, Accounting and Statistics: The Input-Output Framework. *Research Policy* 36 (9): 1388–1403.

Goldstein, J., and Tyfield, D. 2018. Green Keynesianism: Bringing the Entrepreneurial State Back in(to Question)? *Science as Culture* 27 (1): 74–97.

Graham, S., and Marvin, S. 2001. *Splintering Urbanism: Networked Infrastructures, Technological Mobilities and the Urban Condition*. London: Routledge.

Hart, K., ed. 2017. *Money in the Human Economy*. New York: Berghahn.

Hart, K., and Ortiz, H. 2014. The Anthropology of Money and Finance: Between Ethnography and World History. *Annual Review of Anthropology* 43: 465–482.

Haskel, J., and Westlake, S. 2018. *Capitalism without Capital: The Rise of the Intangible Economy*. Princeton, NJ: Princeton University Press.

Helms, S., Liehr-Gobbers, K., and Storck, C., eds. 2011. *Reputation Management*. Berlin: Springer-Verlag.

Ho, K. 2009. *Liquidated: An Ethnography of Wall Street*. Durham, NC: Duke University Press.

Hodgson, G. M. 2003. The Enforcement of Contracts and Property Rights: Constitutive versus Epiphenomenal Conceptions of Law. *International Review of Sociology* 13 (2): 375–391.

Hodgson, G. M. 2014. What Is Capital? Economists and Sociologists Have Changed Its Meaning: Should It Be Changed Back? *Cambridge Journal of Economics* 38 (5): 1063–1086.

Hope, J. 2008. *Biobazaar: The Open Source Revolution and Biotechnology*. Cambridge, MA: Harvard University Press.

Jessop, B. Young, B., and Scherrer, C., eds. 2015. *Financial Cultures and Crisis Dynamics*. London: Routledge.

Joskow, P. 1988. Asset Specificity and the Structure of Vertical Relationships: Empirical Evidence. *Journal of Law, Economics, & Organization* 4 (1): 95–117.

Kamiyama, S., Sheehan, J., and Martinez, C. 2006. Valuation and Exploitation of Intellectual Property. *OECD Science, Technology and Industry Working Papers*, 2006/05. Paris: Organization for Economic Cooperation and Development.

Kang, H. Y. 2015. Patent as Credit: When Intellectual Property Becomes Speculative. *Radical Philosophy* 194: 29–37.

Keynes, J. M. 1936. *The General Theory of Employment, Interest and Money*. London: Macmillan.

Kiely, R. 2018. *The Neoliberal Paradox*. Cheltenham: Edward Elgar.

Konings, M. 2018. *Capital and Time: For a New Critique of Neoliberal Reason*. Stanford, CA: Stanford University Press.

Krippner, G. R. 2011. *Capitalizing on Crisis: The Political Origins of the Rise of Finance*. Cambridge, MA: Harvard University Press.

Langley, P. 2006. The Making of Investor Subjects in Anglo-American Pensions. *Environment and Planning D: Society and Space* 24 (6): 919–934.

Langley, P. 2008. *The Everyday Life of Global Finance: Saving and Borrowing in Anglo-America*. Oxford: Oxford University Press.

Langley, P., and Leyshon, A. 2017. Platform Capitalism: The Intermediation and Capitalisation of Digital Economic Circulation. *Finance and Society* 3 (1): 11–31.

La Porta, R., Lopez de Silanes, F., Shleifer, A., and Vishny, R. W. 1988. Law and Finance. *Journal of Political Economy* 106 (6): 1113–1155.

Larder, N., Sippel, S. R., and Argent, N. 2018. The Redefined Role of Finance in Australian Agriculture. *Australian Geographer* 49 (3): 397–418.

Lépinay, V.-A. 2011. *Codes of Finance: Engineering Derivatives in a Global Bank*. Princeton, NJ: Princeton University Press.

Levidow, L., and Young, R., eds. 1981. *Science, Technology and the Labour Process*. London: CSE Books.

Levy, J. 2014. Accounting for Profit and the History of Capital. *Critical Historical Studies* 1 (2): 171–214.

Levy, J. 2017. Capital as Process and the History of Capitalism. *Business History Review* 91 (3): 483–510.

Leyshon, A., and Thrift, N. 2007. The Capitalization of Almost Everything: The Future of Finance and Capitalism. *Theory, Culture and Society* 24 (7-8): 97–115.

Lezaun, J., and Montgomery, C. 2015. The Pharmaceutical Commons: Sharing and Exclusion in Global Health Drug Development. *Science, Technology and Human Values* 40 (1): 3–29.

Li, T. 2014. What Is Land? Assembling a Resource for Global Investment. *Transactions of the Institute of British Geographers* 39 (4): 589–602.

Li, T. 2017. Rendering Land Investible: Five Notes on Time. *Geoforum* 82: 276–278.

Luque, E. 2001. Whose Knowledge (Economy)? *Social Epistemology* 15 (3): 187–200.

Lyotard, J.-F. 1984. *The Postmodern Condition: A Report on Knowledge*. Manchester: Manchester University Press.

Lyotard, J.-F. 1992. *The Inhuman: Reflections on Time*. Stanford, CA: Stanford University Press.

MacKenzie, D. 2006. *An Engine, Not a Camera: How Financial Models Shape Markets*. Cambridge, MA: MIT Press.

MacKenzie, D. 2009. *Material Markets: How Economic Agents Are Constructed*. Oxford: Oxford University Press.

Marazzi, C. 2011. *Capital and Affects: The Politics of the Language Economy*. Cambridge, MA: MIT Press.

Martin, P. 2015. Commercialising Neurofutures: Promissory Economies, Value Creation and the Making of a New Industry. *BioSocieties* 10 (4): 422–443.

Maurer, B. 2006. The Anthropology of Money. *Annual Review of Anthropology* 35: 15–36.

Mazzucato, M. 2013. *The Entrepreneurial State: Debunking Public vs. Private Sector Myths*. London: Anthem Press.

Mazzucato, M. 2018. *The Value of Everything: Making and Taking in the Global Economy*. London: Allen Lane.

McFall, L. 2014. *Devising Consumption: Cultural Economies of Insurance, Credit and Spending*. London: Routledge.

McGoey, L. 2015. *No Such Thing as a Free Gift: The Gates Foundation and the Price of Philanthropy*. London: Verso.

Mennicken, A., and Lodge, M. 2015. Quantification, Administrative Capacity and Democracy. *Risk & Regulation* 30: 6–9.

Mennicken, A., and Muniesa, F. 2017. Governing through Value: Public Service and the Asset Rationale. *Risk & Regulation* 34: 9–10.

Miller, P. 1991. Accounting Innovation beyond the Enterprise: Problematizing Investment Decisions and Programming Economic Growth in the UK in the 1960s. *Accounting, Organizations and Society* 16 (8): 733–762.

Mirowski, P. 2011. *Science-Mart: Privatizing American Science*. Cambridge, MA: Harvard University Press.

Mirowski, P. 2012. The Modern Commercialization of Science as a Passel of Ponzi Schemes. *Social Epistemology* 26 (3-4): 285–310.

Moore, J. W. 2015. *Capitalism in the Web of Life: Ecology and the Accumulation of Capital*. London: Verso.

Moulier Boutang, Y. 2011. *Cognitive Capitalism*. Cambridge: Polity.

Muniesa, F. 2012. A Flank Movement in the Understanding of Valuation. *Sociological Review* 59 (s2): 24–38.

Muniesa, F. 2016. The Matter of Business. In *Materiality, Rules and Regulation: New Trends in Management and Organization Studies*, edited by F.-X. de Vaujany, N. Mitev, G. F. Lanzara, and A. Mukherjee, 121–126. Basingstoke: Palgrave Macmillan.

Muniesa, F. 2017. On the Political Vernaculars of Value Creation. *Science as Culture* 26 (4): 445–454.

Muniesa, F., Doganova, L., Ortiz, H., Pina-Stranger, A., Paterson, F., Bourgoin, A., Ehrenstein, V., Juven, P-A., Pontille, D., Saraç-Lesavre, B., and Yon, G. 2017. *Capitalization: A Cultural Guide*. Paris: Presses des Mines.

Muniesa, F., and Linhardt, D. 2011. Trials of Explicitness in the Implementation of Public Management Reform. *Critical Perspectives on Accounting* 22 (6): 550–566.

Neyland, D. 2018. On the Transformation of Children at-Risk into an Investment Proposition: A Study of Social Impact Bonds as an Anti-market Device. *The Sociological Review* 66 (3): 492–510.

Nitzan, J., and Bichler, S. 2009. *Capital as Power: A Study of Order and Creorder*. London: Routledge.

O'Brien, P., and Pike, A. 2015. The Governance of Local Infrastructure Funding and Financing. *Infrastructure Complexity* 2: 1–9.

OECD. 2008. *Knowledge Markets in the Life Sciences*. Paris: Organization for Economic Cooperation and Development.

OECD. 2010. *Handbook on Deriving Capital Measures of Intellectual Property Products*. Paris: Organization for Economic Cooperation and Development.

OECD. 2012. *New Sources of Growth: Knowledge-based Capital Driving Investment and Productivity in the 21st Century*. Paris: Organization for Economic Cooperation and Development.

O'Neill, P. 2013. The Financialisation of Infrastructure: The Role of Categorisation and Property Relations. *Cambridge Journal of Regions, Economy and Society* 6 (3): 441–454.

Orléan, A. 2014. *The Empire of Value: A New Foundation for Economics*. Cambridge, MA: MIT Press.

Orr, R. J. 2007. The Rise of Infra Funds. *PFI Global Infrastructure Report 2007*.

Ortiz, H. 2013. Financial Value: Economic, Moral, Political, Global. *Hau: Journal of Ethnographic Theory* 3 (1): 64–79.

Ortiz, H. 2014. The Limits of Financial Imagination: Free Investors, Efficient Markets, and Crisis. *American Anthropologist* 116 (1): 38–50.

Ouma, S. 2019. This Can('t) Be an Asset Class: The World of Money Management, "Society," and the Contested Morality of Farmland Investments. *Environment and Planning A: Economy and Space*, doi:10.1177/0308518X18790051.

Owen, G., and Hopkins, M. 2016. *Science, the State, and the City*. Oxford: Oxford University Press.

Palan, R. 2013. The Financial Crisis and Intangible Value. *Capital & Class* 37 (1): 65–77.

Parthasarathy, S. 2017. *Patent Politics: Life Forms, Markets, and the Public Interest in the United States and Europe*. Chicago: Chicago University Press.

Perzanowski, A., and Schultz, J. 2016. *The End of Ownership: Personal Property in the Digital Economy*. Cambridge, MA: MIT Press.

Petersen A., and Krisjansen, I. 2015. Assembling "the Bioeconomy": Exploiting the Power of the Promissory Life Sciences. *Journal of Sociology* 51 (1): 28–46.

Piketty, T. 2014. *Capital in the 21st Century*. Cambridge, MA: Belknap Press.

Pisano, G. 1991. The Governance of Innovation: Vertical Integration and Collaborative Arrangements in the Biotechnology Industry. *Research Policy* 20 (3): 237–249.

Pistor K. 2013. A Legal Theory of Finance. *Journal of Comparative Economics* 41 (2): 315–330.

Pistor, K. 2019. *The Code of Capital: How the Law Creates Wealth and Inequality*. Princeton, NJ: Princeton University Press.

Pitelis, C., and Teece, D. 2009. The (New) Nature and Essence of the Firm. *European Management Review* 6: 5–15.

Polanyi, K. 1944. *The Great Transformation: The Political and Economic Origins of Our Time*. New York: Farrar & Rinehart.

Porter, T. M. 1995. *Trust in Numbers: The Pursuit of Objectivity in Science and Public Life*. Princeton, NJ: Princeton University Press.

Pryke, M., and Du Gay, P. 2007. Take an Issue: Cultural Economy and Finance. *Economy and Society* 36 (3): 339–354.

Rifkin, J. 2014. *The Zero Marginal Cost Society: The Internet of Things, the Collaborative Commons, and the Eclipse of Capitalism*. New York: Palgrave Macmillan.

Riordan, M., and Williamson, O. E. 1985. Asset Specificity and Economic Organization. *International Journal of Industrial Organization* 3: 365–378.

Risch, M. 2013. Patent Portfolios as Securities. *Duke Law Journal* 63: 89–154.

Robson, K. 1993. Governing Science and Economic Growth at a Distance: Accounting Representation and the Management of Research and Development. *Economy and Society* 22 (4): 461–481.

Robson, K. 1994. Connecting Science to the Economic: Accounting Calculation and the Visibility of Research and Development. *Science in Context* 7 (3): 497–514.

Sadowski, J. 2019. When Data Is Capital: Datafication, Accumulation, and Extraction. *Big Data & Society*, doi:10.1177/2053951718820549.

Saraç-Lesavre, B. 2017. In Search of an Assessment of the Future: The Case of the US Nuclear Waste Programme. In *Socioeconomic Evaluation of Megaprojects*, edited by M. Lehtonen, P.-B. Joly, and L. Aparicio, 25–43. London: Routledge.

Saraç-Lesavre, B. 2018. Budgeting the Future: Negotiating the Values of a Contemporary Liberal Democracy. In *Materiality and Managerial Techniques: New Perspectives on Organizations, Artefacts and Practices*, edited by N. Mitev, A. Morgan-Thomas, P. Lorino, F.-X. de Vaujany, and Y. Nama, 57–76. New York: Palgrave Macmillan.

Schwartz, H. M. 2016. Wealth and Secular Stagnation: The Role of Industrial Organization and Intellectual Property Rights. *Russell Sage Foundation Journal of the Social Sciences* 2 (2): 226–249.

Schwartz, H. M. 2017. Club Goods, Intellectual Property Rights, and Profitability in the Information Economy. *Business and Politics* 19 (2): 191–214.

Shore, C., and Wright, S. 2000. Coercive Accountability: The Rise of Audit Culture in Higher Education. In *Audit Cultures: Anthropological Studies in Accountability, Ethics and the Academy*, edited by M. Strathern, 57–89. London: Routledge.

Springer, S., Birch, K., and MacLeavy, J., eds. 2016. *The Handbook of Neoliberalism*. London: Routledge.

Srnicek, N. 2017. *Platform Capitalism*. Cambridge: Polity.

Star, S. L. 1999. The Ethnography of Infrastructure. *American Behavioral Scientist* 43 (3): 377–391.

Sum, N. L., and Jessop, B. 2013. *Towards a Cultural Political Economy: Putting Culture in its Place in Political Economy*. Cheltenham: Edward Elgar.

Sunder Rajan, K. 2006. *Biocapital: The Constitution of Postgenomic Life*. Durham, NC: Duke University Press.

Sunder Rajan, K. 2017. *Pharmocracy: Value, Politics, and Knowledge in Global Biomedicine*. Durham, NC: Duke University Press.

Toms, J. S. 2010. Calculating Profit: A Historical Perspective on the Development of Capitalism. *Accounting, Organizations and Society* 35 (2): 205–221.

Tooze, A. 2018. *Crashed: How a Decade of Financial Crises Changed the World*. New York: Viking.

Van der Zwan, N. 2014. Making Sense of Financialization. *Socio-Economic Review* 12 (1): 99–129.

Veblen, T. 1908a. On the Nature of Capital: Investment, Intangible Assets and the Pecuniary Magnate. *Journal of Economics* 22 (4): 517–542.

Veblen, T. 1908b. On the Nature of Capital: Investment, Intangible Assets and the Pecuniary Magnate. *Journal of Economics* 23 (1): 104–136.

Vezyridis, P., and Timmons, S. 2017. Understanding the care.data Conundrum: New Information Flows for Economic Growth. *Big Data and Society* 4 (1): 1–12.

Ward, C., and Swyngedouw, E. 2018. Neoliberalisation from the Ground Up: Insurgent Capital, Regional Struggle, and the Assetisation of Land. *Antipode* 50 (4): 1077–1097.

Williamson, O. E. 1979. Transaction-Cost Economics: The Governance of Contractual Relations. *Journal of Law and Economics* 22 (2): 233–261.

World Economic Forum. 2011. *Personal Data: The Emergence of a New Asset Class*. Geneva: World Economic Forum.

Zaloom, C. 2006. *Out of the Pits: Traders and Technology from Chicago to London*. Chicago: University of Chicago Press.

Zeller, C. 2008. From the Gene to the Globe: Extracting Rents Based on Intellectual Property Monopolies. *Review of International Political Economy* 15 (1): 86–115.

I Turning Knowledge into Assets

2 Patents as Assets: Intellectual Property Rights as Market Subjects and Objects

Hyo Yoon Kang

Introduction

Patents occupy a hybrid node in the entanglement of science, technology, and finance within capitalist economy and law. A patent, an intellectual property right creating a monopoly of twenty years, contains different proprietary modes in which an invention's potential may be materialized in social relations: via appropriation, possession, commodification, and assetization. These modes may not necessarily always overlap. This chapter describes and problematizes a specific turn to assetization that patents have taken: the transfiguration of patents into speculative financial assets. In light of the scholarship about the marketization and financialization of sciences (Nelkin 1984; Mirowski 2011; Birch 2017) and cultural studies of capitalization processes (Muniesa et al. 2017), I extend the question of patent value (Kang 2015) to examine the practices and mechanisms of valuation by which patents—legal property rights—are transformed into assets. I delineate the different ways in which patents are valued and acted upon as financial assets, which are premised on layers of legal and financial abstraction. Whereas it is well known that patents commodify, alienate, and eclipse their original referents—the inventions (Strathern 1999)—the analysis here shows that patents are enacted as real options in valuation practices and have been used as instruments of financial hedging. As a result, I argue that law itself is turned into a speculative financial asset.

The chapter focuses on the novel forms and practices that have turned the legal form of patents into speculative financial assets rather than offering an analysis of patents as legal techniques of commodification and monopoly rent-seeking. From the perspective of intellectual property law scholarship, the often-voiced criticisms against patents, that they exclude production and labor from ownership and limit access to inventions through exclusionary practices and unfettered pricing, are not novel. Modern patents are monopoly rights that have been created to legally sanction such practices and effects. These criticisms nonetheless put the underlying justification of the patent system into question. This rests on the belief that a temporary monopoly promotes progress (e.g., US Constitution, Art. I, Sect. 8, Clause 8). A patent is expressly intended for the creation of a monopolistic market in order to reward inventive activities. This in turn is believed to generate more innovation in the long run (Schumpeter 1976). It is therefore neither new nor surprising that patents serve as mechanisms of generating monopoly rents on innovations (Birch 2017). That has precisely been the legislative intent of modern patent monopolies.

Intellectual property rights are often presumed to be valuable assets in the "knowledge economy" (UK Department for Business, Innovation and Skills 2016) or "intangible economy" (Haskel and Westlake 2018). Such an equation between patents as intangible assets and their value is assumed rather than explicated, or the precise nature of such a relationship queried. If patents are presumed to be valuable assets, then the question is what specific forms and material practices facilitate and enact patents' assetization. In other words, what kind of assets *are* patents at this moment of financial capitalism, and what kind of understandings of market and law do the concrete assetization practices reveal? In common language, interdisciplinary literature of science and technology studies (STS), as well as in balance-sheet accounting rules, intellectual property rights are defined as intangible assets. However, it is not evident that patents are assets, or at least valuable assets, unless they are enlisted in specific modalities of value or acted upon in specific ways. The question of patents as assets is not only a question of nomenclature or a presumed derivative of contemporary political economy. But it entails a specific examination of the interaction of patents, which are abstract, generalized legal forms (Boltanski and Thevenot 2006, 8), with techniques of valuation and practices of transaction that enact their potential value. The effectiveness of patents as assets is utmost contingent on many legal formal requirements of representations and strategies (Menell 2018). It is also dependent on the availability and effectiveness of enforcement through litigation and remedies (Lanjouw and Schankerman 2001). Yet patents are turned into assets through practices and specific knowledge techniques that are outside the "legal" realm. The aim of this chapter is to investigate some of these specific sites of patents' valuation and their financial assetization.

It is helpful to tread carefully around vocabularies in order to differentiate what patents are and what they do. The following words in particular often become conflated in discussions about intellectual property: originality, creativity, knowledge, invention, innovation, patent, property, tangibility, and intangibility. The inherent hybrid character of these terms and concepts becomes purified in law (Latour 1993), and such rhetorical separations give rise to distinct realities, not dissimilar to the way in which economic models and business education literature perform certain kinds of economies (Muniesa 2014). For analytical differentiation, I hold on to the legal definition of a patent as an intellectual property right, which is essentially an abstract legal form exerting practical effects. This is not because I believe that the legal self-definition is a comprehensive or truthful one but because holding on to it allows me to distinguish and trace the legal form's instability and contingency by interpreting it through the lenses of social theory and STS scholarship and delineating its various shapes as a result. Being attentive to these arbitrary, disciplinary, fictional, and material distinctions affords a better vision of what is at stake when patents are turned into and enacted as financial assets. It also allows for a differentiation between a specific financial logic of assetization in contrast to other modalities in which patents act as assets.

"When Wealth Lives Mainly in Intellectual Property": Patent as Privilege, Property, Commodity, Asset

Patents were initially a privilege: they were granted not as a right but as a favor by the sovereign. One of the earliest sovereign patents was granted as a commercial monopoly for a useful invention in Venice in 1469 (Kostylo 2008). Patents were not transferable and were only intended for the person in use. An ensemble of abstract legal norms, infrastructure networks, and practices have been built in order to maintain and normalize the modern category of intellectual property as a legal property right, which can be possessed, used, transferred, exchanged, and sold. But this had not always been so, and there is no inevitable sense why a patent should be a property right rather than an exception to the general rule of no special protection. There have been precedents of alternative intellectual property arrangements or a lack thereof: for example, the Netherlands abolished its patent system in 1869 and did not reintroduce it until 1912 (Moser 2005). Not so long ago, there have also been differences in national patent laws prior to the coming to force of the international system of intellectual property law via the Trade Related Aspects of Intellectual Property Rights (TRIPS) Agreement, backed up by the World Trade Organization. Historically Germany, as well India until 2005 when TRIPS had to be implemented, did not allow patents to be granted on products and thus promoted other ways to invent around an existing patent.[1]

The modern rationale behind the creation of legal monopoly rights, such as patents, rests on the temporal postulates of *ex post* (reward for labor and investment expended) and *ex ante* (belief in incentivization of innovation by temporally restricted monopoly rights) effects of patent law. They incorporate the belief that the public encouragement of temporally limited private rent-seeking through patents is beneficial to the public. Much of the economic, policy-oriented, and economic history scholarship has been devoted to the evaluation of the long-term economic consequences of short-term monopolistic practices (Dasgupta and Stiglitz 1988; see Moser 2013 for an overview). Beyond the narrative of fostering domestic innovation for a vision of public good based on an economically competitive nation-state, economic historians have shown that much of the value of patents has been strategic and international in scope since the Paris Convention of 1883, which was the first attempt of an international coordination of patent policy (Bilir et al. 2011; Ricketson 2015). Patents remain prominent instruments of trade that can facilitate a commodity's market access or hindrance (Ryan 1998; Drahos and Braithwaite 2002). Although they are regarded as an alternative instrument of trade policy to tariffs, the effectiveness of patents as a trade instrument depends on their value and valuation (CBC News 2018). Recently some legal scholars have argued that the incentive-rationale of commodities trading of the international patent regime under TRIPS is increasingly coming into conflict with investment treaties in which patents are treated as investment assets rather than commodities (Dreyfuss and Frankel 2015). Such predominant ideologies and treatments of patent value are premised on the belief in free market and trade that separate asset value from the conditions of its production.[2]

These studies of patents as instruments of political economy may be helpful for understanding its current multiple jurisdictional layers and legal technicalities; however, they also risk reifying and ascribing a force to patents which they do not always have. Legal materialist approaches to the study of intellectual property rights have emphasized the contingency of these rights and legal forms, rather than understanding them as abstract rights that have uniform realities. As the essence of intangible knowledge has not been easy to capture, intellectual property law has instituted a web of legal forms (abstract property rights administered through a registration system, specific format and writing), uses (disclosure, ownership, possession, non-use, exclusion), and exchanges (through sale or licenses) to outline and thereby define the products of intellectual labor. Physical boundaries in land or the exteriors of buildings are visible but knowledge boundaries less so. The categories of intellectual properties in intangibles have been utmost materially choreographed and stabilized (Pottage and Sherman 2010; Kang 2012; Bellido and Kang 2016; Kang 2019). In the context of patent law, a patent's proprietary boundaries are determined by the textual boundaries of claims and the overall composition and intertextuality of a patent document (Myers 2005). The evolution of the modern patent law system of registration, disclosure, examination went hand in hand with the creation of a considerable administrative structure that established paper trails, textual rules, bureaucratic procedures, which were based on classification of knowledge (Kang 2012).[3]

Despite the contingent nature of intellectual property rights, a certain naturalized understanding of intellectual property spread to and was adopted by other domains of knowledge and values, which ascribed more and sometimes too little value to patents than they arguably actually had. This was not only a material effect of legal rhetoric (Edelman 1979) which eclipsed the original object of representation, the invention (Strathern 1999). The conflation of patents with value accelerated with different understandings of intellectual property as, on the one hand, a legal Schumpeterian instrument in a capitalist economy driven by innovation, and on the other hand, a problem of justice voiced by critical legal scholars who were concerned with the inequitable effects of TRIPS in particular, as well as scholars in anthropology and STS in the political economy of the biosciences.

In the body of interdisciplinary and critical scholarship on patents found in law, anthropology, and STS, there have been numerous analyses and critiques of patent law as the central legal technique of commodification within the so-called knowledge economy. These concerns do not feature in formalist patent law scholarship because, from a doctrinal point of view, patents do not give ownership in knowledge. Patents cannot be granted for discoveries, principles of nature, or knowledge, but only for novel and useful embodiments of an inventive essence or for an inventive process. In an ideal world, the quality of the patent examination process would be high enough to adequately assess an application according to the formal and substantive patentability requirements: sufficient and enabling disclosure, novelty, non-obviousness, and utility. Yet as the 2013 US Supreme Court case of *Association for Molecular Pathology v. Myriad Genetics Inc.*[4] has demonstrated, the boundaries of conceptual dichotomies underlying patent law's doctrinal principles, such as nature/artifice and persons/things, are stretched to the limits of their meaning in the context of biotechnology.

A reoccurring trope of concern has been the commodification of nature or knowledge, which ought not to be privately owned or enclosed, as well as the critique of the severance between the material circumstances of knowledge production and the abstract legal form of a patent. Critiques of legal techniques of commodification through intellectual property law have often employed the notions of public domain, commons and openness (Heller and Eisenberg 1998; Biagioli 2009; Boyle 2010). Patents have been characterized as turning nature and culture into market commodities (Coombe 1998) and establishing a barrier or blockage in the access to knowledge and medicines (Krikorian and Kapczynski 2010; Cassier and Correa 2014). In more concrete settings, anthropologists have depicted a complex picture of discourses and practices of intellectual property which complicate established dichotomies between private vs. public domains and openness vs. enclosure (Hayden 2003, Aragon and Leach 2008; Kelty 2008; Peterson 2014). Anthropological studies of biotechnology have highlighted the links between tropes of speculative finance capitalism and the intellectual property rights which go beyond a commodity market logic (Sunder Rajan 2006; Fortun 2008).

In contrast to such detailed critiques, the belief that "wealth mainly lives in intellectual property" (*Financial Times* 2017) has become a common parlance in management literature (Drucker 1969) and has been taken up by government and international organizations (UK Intellectual Property Office 2017). Intellectual property, particularly a patent, is seen as an asset in a Schumpeterian understanding of the "knowledge economy" driven by a chain of disruptive innovations. Here intellectual property is understood self-referentially as both a driver of value creation as well as an indicator of innovation, the latter positively interpreted to be a determinant of economic growth and therefore assumed to be of value.

Such a circular view can also be found in discourses about the role of the university and the characterization of research and teaching in terms of their impact on the "knowledge economy." The overlap between intellectual property and knowledge is assumed and not unpacked. An understanding of intellectual property as an asset, however, would need to be embedded in a diagnosis of the contemporary meaning of knowledge itself and its status within present configurations of economy (Raunig 2013; Lazzarato 2014). From the perspective of critical legal scholarship, the acquisition of facticity and naturalization of intellectual property are problematic because they ignore its constructed and contingent nature. Nonetheless this insight does not deny intellectual property's rhetorical performativity and invocation of certain realities. Patent offices themselves call upon such a nexus of "innovation—economic growth—intellectual property" by reference to future potentiality: "Innovation fuels economic growth. There is evidence to show that more innovative markets are the ones that grow. This is true across the whole economy or individual industries, with the businesses in them measurably more productive. IP is important for innovation" (UK Intellectual Property Office 2017). Such a claim reflects a self-referential tautology: intellectual property drives innovation, innovation is necessary to create growth in the modern knowledge economy, and innovation is measured by the patent information, such as the number of patent citations, number of applications, and patent renewals.

Patent information has become valuable as an economic unit itself, for example, patent raw data are priced by the European Patent Office (EPO 2017). Economists have adopted the narrative of patents as assets by using patent data as an indicator of innovation (Griliches et al. 1987; Griliches 1990), as have scientometric methods (Leydesdorff 2004), and the measurement of the effects of academic research (Jaffe 1989). Qualitative survey data, however, has yielded a more complicated picture, which shows that the incentives for inventions are not always of a monetary nature (European Commission 2005). Nor has economic scholarship always taken patent quality into account, which can only be assessed by opening up the patent document through textual interpretation rather than by patent data statistics. As indicators of patent value, economists have employed quantitative factors such as patent renewal (Pakes 1986; Lanjouw et al. 1998), rate of litigation (Lanjouw and Schankerman 1997), or patent citations (Trajtenberg 2002). The predominant strand of patent law scholarship employing law and economics methods has been inconclusive in its assessment of the overall benefits of the patent system for fostering innovation, but it identified many of what it regarded as the system's dysfunctionalities and pushed for its reform (Jaffe and Lerner 2006; Bessen and Meurer 2008; Burk and Lemley 2009). Patents continue to be regarded as something inherently valuable, although around 75 percent of patents are not useful in the sense that they are licensed out, referred to, or used (European Commission 2005). The above analysis reflects how the understanding of patents has changed from being primarily understood as assets rewarding inventors to key infrastructural assets for and within the "knowledge economy" (Kamiyama et al. 2006), a term that lacks a clear definition. The following sections describe the modes and practices in which patents act and are transacted as assets: patent portfolios, financial valuations, and market hedges.

Patent Portfolios: When the Intangible Is the Real

Patents contain distinctive modalities of value: they can denote a retrospective credit for work, act as currencies of credence, and serve as a financial security or an asset (Kang 2015). As industrial and business strategy, patents have been studied as strategic assets for the industrialization of knowledge, often crossing the dichotomy of pure and applied sciences, particularly in the life sciences and chemistry since the nineteenth century and used for gaining international competitiveness (Gaudilliere and Loewy 1998; Homburg et al. 1998; Steen 2000; Cassier 2005; Galvez-Behar 2016). The value of patents as assets derives from the projected exchange value of the patent as a property right rather than the actual or anticipated use value of the patented invention. The latter would be based on the terms of license, the individual contents of which cover conditions of use (exclusive, non-exclusive, time frame, one-off fee payment, or continuous royalty, etc.) and are negotiated within the bounds of contract law.

In contrast, the logic of patents as financial assets takes as its object the legal form of *property itself*: What is valued are not the potential licenses and market size for the patented invention but the future return expected on the intellectual property right as a forward-looking investment vehicle (Risch 2013). In such a forward-looking financial logic, the black box of the legal form, the patent, is seldom opened up or evaluated in terms of its relative strength and quality. The investment logic expressly disconnects patent ownership from labor or the actual use of the patented invention by others. What matters in terms of a patent's financial asset value is the property right itself as a financial vehicle rather than the commercial potentiality of an invention.[5] In controversies around "patent trolls," nonpracticing entities that buy patents with a view of obtaining a settlement or damages through litigation against alleged patent infringers, arguments against "free-riding" and the inequity of deriving profit from someone else's inventive labor have been voiced (Bessen and Meurer 2008). Treating patents as investment assets conflicts with the patent system's premise and justification that there is a balance to be struck between monopoly rights and the public interest in "the progress of sciences and useful arts." This utilitarian consideration is often referred to as the "patent bargain." There is, however, no discernible alignment between patent assetization and scientific progress; in fact, the financial assetization of patents seems to run counter to it (Tucker 2014). In order to ascertain whether treating patents as investments would be compatible with the legal narrative of a patent bargain, it would be necessary to assess their effects on the primary economy of production in terms of patent litigations initiated by the nonpracticing entities, the size of the litigated companies, and the damages awarded by the courts.

Recent transactions by operating companies indicate, however, that the so-called secondary or derivative market in patents *is* the primary market. The patent portfolio wars in the information and communication technology sector have radically destabilized if not completely dissolved the equation between primary market with tangible goods and secondary (or virtual) market with intangible products. According to Ocean Tomo, which tracks the value of intangibles in the stock market, the value of intangibles has overtaken the value of tangibles within the overall S&P market capitalization in the US. In 1975 more than 80 percent of corporate value reflected in the S&P 500 was attributed to tangible assets, while intangible assets comprised less than 20 percent of market capitalization. As of 2017, the ratio of tangible to intangible assets has inverted—nearly 80 percent of corporate value resided in intangible assets.[6] Patent portfolios have become separate, distinct assets, the value of which as a whole is deemed worth more than the sum of its constitutive parts. As Wagner and Parchomosky (2005) point out, companies will seek large quantities of patents as a portfolio rather than evaluating individual patents' actual worth.

Past high-profile transactions in the field of information technology have indeed valued patent portfolios as more desirable assets than the physical assets on which they were based. The premium placed on acquiring property rights, which gives power of control rather than production capacity, might be due to the patent thickets that have particularly been acute in technologies relating to smartphones and that have led companies to work around them through patent pools (Barnett 2014). In this particular sector, patents have been purchased as strategic assets, as arsenals in a 'war' of mutual patent portfolio containment directed at competitors. In 2011, Nortel, a large Canadian telecoms equipment manufacturer, filed for bankruptcy upon which a patent auction of its patent portfolio raised $4.5 billion paid by a consortium of companies including Apple and Microsoft. Or when Google bought Motorola's patents in 2011 for $12.5 billion, it was mainly interested in Motorola's phone technology patents, and sold off the smartphone business to Lenovo in 2014 for $3 billion, effectively valuing Motorola's patents three times more valuable than the physical business. Another example, the demise of the venerable analogue film company Eastman Kodak was made even more poignant by the fact that the remaining value of the company was predominantly based on its intellectual property portfolio, particularly its patents portfolio, rather than its physical assets. It was reported that between 2008 and 2011 almost $2 billion of Kodak's revenue was generated through licensing fees, royalties, and intellectual property related litigations and settlements. In 2012, 1,101 of its digital imaging patents were sold to a consortium of bidders, that included Google, Apple, Facebook, and Samsung, for $525 million. The consortium was led by Intellectual Ventures and RPX, two prominent nonpracticing entities. Contrary to the perception that operating companies are at loggerheads with nonpracticing companies, the association and mixture of companies shows complex webs of overlapping interests and people between these companies, so much so that so-called operating companies also act as patent funds, patent brokers, and nonpracticing entities (patent trolls).[7] For example, the founder of Intellectual Ventures, Nathan Myhrvold, had been the chief strategist and chief technology officer at Microsoft. It is a chicken and egg question whether the patent arsenal buildup is a defensive or an aggressive strategy employed by operating companies in order to defend themselves against patent lawsuits from both their competitors, as well as trolls. What these past transactions indicate is that patents as property rights are the coveted primary assets rather than being merely seen as means of commodification.

Accounting for Patent Value versus Modeling for Patent Value

As Power (1992) and Sherman and Power (1994) have shown in the context of brand value accounting, a quantitative valuation of intellectual property involves a rematerialization of a legal potential (of a trademark or a patent) in a specific social context. The making of patents' "order of worth" (Boltanski and Thevenot 2006) has been accompanied by a call to better account for the value of patents. Quantitative models estimate the value of patents as intangible assets for accounting or investment purposes.[8] Much of the reality of patents as financial assets is numerically articulated as the return on investment that is modeled on an Excel spreadsheet and acted upon in financial transactions. This section takes a look at some of the most commonly used quantitative patent valuation methodologies.

Patents have acted as collateral for capitalization, assets in a balance sheet, and as investment vehicles.[9] These assetization practices require ascertaining a patent's monetary value. Different valuation methodologies can be used, depending on the purpose of the valuation, which may be balance sheet accounting, market transaction, or investment. The methodologies are, in turn, correspondingly based on parameters of cost, income, or market; or it can also involve real option pricing models. The last methodology could either employ the Black-Scholes equation or the binomial option pricing model (see Gilbert, this volume).

The valuation of intangible assets for the balance sheet is often much below a market valuation of a company as it is reflected in the share price. In 1999, a PricewaterhouseCoopers report listed two main methods of valuing intellectual property that it deemed suitable: income or cost-based methods. It excluded market-based valuation for patents because that would presuppose an existing market with comparables and the availability of sufficient public information. Given that novelty is one of the legal prerequisites for obtaining a patent, very novel patented inventions often do not have market comparables. Also as licenses are private contracts, there is often a lack of reliable licensing information, particularly for unlisted companies. For these reasons, a market-based valuation approach was deemed insufficiently reliable. An income-based valuation was only suggested for active patents that already generated cash. For inactive patents with potential for future use, the 1999 report suggested a cost-based approach to patent valuation, which would be based on the projected cost of replacement—an exercise that would mainly capture the cost of filing for a patent but not an invention's future potential value. What is remarkable about the 1999 report is that its valuation methods categorize intellectual property as a commodity asset rather than as a financial asset. Intellectual property was to be valued on the basis of past earnings, excluding forward-looking estimates. As a result, it yielded conservative valuations.

Such a cautious approach to financial accounting stands in contrast to the forward-looking methods employed for company valuations or license valuations, often used for biopharmaceutical patents (see Roy, this volume). A valuation textbook puts it this way: "Start with the obvious.... Intangible assets are worth a lot and accountants don't do a good job in assessing their value" (Damodaran 2006, 2). Most commonly, if a patent was already cash-generating, the discounted cash flow method would be used, which is based on projected income through licensing, royalties, or sale during the patent's lifespan and discounts it with a factor taken from patent holder's industry peers, hypothetically reflecting the riskiness of these future cash flows and the anticipated required returns on capital employed. This mode of valuation is speculative, especially before regulatory approval of a medicine or therapy, but companies use it widely for capital raising or loans. In 2012, the discounted cash flow method was reported to be the most commonly used method by so-called IP brokers (Escoffier and Kasznik 2012).

Reflecting the future-oriented and speculative nature of patents as property forms, patents are increasingly valued using option pricing models. These have been seen as particularly apt for valuing patents that have been granted and do not generate cash flows—or not yet, but may potentially do so in the future. Valuation textbooks suggest to apply real option methods, such as the Black-Scholes model or binomial option pricing model, with a preference for the binomial option when asset prices are not steady (Damodaran 2006). The valuation of patents as real options implies that a patent is predominantly understood as a forward-looking, speculative asset rather than as an embodiment of a commodity value. What is being valued as a real option is not the worth of a patented invention as a commodity, but the shell of proprietary right, the patent, which acts as a vehicle that transports a monopoly right into its potential market futures.

Market in Patent Monetization versus Market in Patents as Financial Assets

Together with these different modes of patent valuation, two types of markets have emerged with distinct operating rationales (MacKenzie 2006). There are first-order commodity markets in patent licenses, which are priced based on an estimation of future income stream of a patent (Gu and Lev 2008). These are markets aimed at the "monetization" of existing patents or patent portfolios, the value of which is otherwise seen to be idling around. They are either initiated by large patent holders themselves (e.g., IBM, Philips) or offered via an intermediary platform (e.g., the Intellectual Asset Management marketplace). The sales of patent portfolios as described earlier are examples of monetization, as the CEO and chairman of Kodak remarked after the patent sale: "This monetization of patents is another major milestone toward successful emergence" (*New York Times* 2012).

Differently from these monetization practices, I identify an additional kind of derivative market, which could be called second-order patent markets. These value patents on the basis of legal solidity and probability of winning adjudicatory disputes. The latter are typical of the patent troll business model of litigation threats and challenges of patent validity via an *inter partes* review at the US Patent and Trademarks Office. The difference between this kind of second-order market to the market in patent monetization is that the former identifies and values patents as financial investments (e.g., in patent portfolio funds, which treat patents as yields rather than commodities).[10] Patents are also seen as assets for hedging and mediating risk. For example, the UK Intellectual Property Office webpage portrays intellectual property as a risk management and assetization strategy in order to maintain psychological or affective market "confidence" on a "fair" return on investment: "The system of interconnected IP rights, patents, trademarks, designs and copyright, reduces the risk of investing in innovation by ensuring that the results can be commercially exploited by the owner and protected from exploitation by others. With good IP protection, innovators can be confident that they can make a fair return on their investment" (UK Intellectual Property Office 2017). Here, patents are characterized as affective assets in the financial market that extends beyond the original commodity market for an invention.

The difference between primary and secondary markets in patents as assets furthermore gives rise to different mechanisms of capital accumulation. In the primary markets for patents, the price of a patent license can be estimated by a discounted cash flow model. The reason for licensing-out a patent is based on a profit rationale which assumes that the license might yield a stable cash flow for a specified period for the licensor, or at least there is a potential for it to do so. The licensee would also have negotiated the use right—be it exclusive or not—with the view that some kind of business or economic benefit would be derived from it. The profit from royalties or license fees normally takes the form of rent: either in the form of a cut of the profits derived from the use of the patent, or a one-off license fee.

In contrast, the secondary market in patents is driven by price arbitrage, devoid of extracting rent or any other reference to the object of property right. The profit stemming from a secondary market does not necessarily have a link to the business of invention itself. The value of patents is based on arbitraging price margins, assessing the strength of legal claims language in the patent document, and speculating on the ability of the legal system to cope with the workload, courts' interpretive inclinations, as well as their willingness to enforce the law. This is the business model of nonpracticing entities (patent trolls), which treat the legal property form as an asset of speculation rather than as vehicles of rent extraction via commodification. The effect of such an arbitrage is twofold: first, the specificity of inventions becomes less relevant to the creation of financial value; and second, the value is predominantly based on the legal form of the patent as a financial vehicle. The financial assetization of patents transforms abstract intellectual property rights into speculative investment vehicles.

Is the Turn to Assets New? Where Patents Act as Assets and Where They Don't

One of the common criticisms about nonpracticing entities has been that they profit from other people's efforts, reaping benefits of what others have sown. But that is exactly the point of an investment vehicle or security (Risch

2013). Knowledge products, such as inventions, are commonly regarded as being less alienable than other objects of property and are seen to possess a different quality than land or tangible objects. If one can speak of alienating knowledge at all, it could denote a lack of recognition through plagiarism (Biagioli 2014) or misattribution of credit by wrongful copying, but not because knowledge can be consumed away, depleted, or is rivalrous (Heller and Eisenberg 1998). Perhaps that is why, unlike in some other branches of property, patent holders are not expected to behave like intermediaries, such as real estate agents or asset managers. They are expected to act like owner-occupiers of a house rather than buy-to-let investors. Yet, despite the myth of the inventive genius (Israel 2000; Bracha 2005), patent law has long separated property ownership (including use and exchange) from inventive labor (Fisk 2009). From a legal technical point of view, there is no formal and ontological difference between real and intellectual properties, and indeed many of intellectual property law's materializations are governed by the same epistemological forms and material techniques, such as registration, documentation, for the inscription of proprietary boundaries as in the other branches of property law (Bhandar 2017).

The monetization and assetization of property rights are not new: trusts are as old as the English system of equity. Objects of real property, such as land and buildings, underlie a complex legal web of leases and licenses, as well as being able to be bundled into abstracted financial parcels such as real estate investment trusts (REITs). As there are secondary and virtual markets in real estate funds—a process of assetization based on property rights in physical objects—similarly, derivative markets in patents have grown and have become more visible. In the US over the last ten years, nonpracticing entities have been characterized as intermediaries, brokers, or middlemen (Hagiu and Yoffe 2013). The founder of the RPX, a patent risk management service, who acted as the former Goldman Sachs chief IP counsel, has described the aim of such intermediaries to "realize value" of dormant patents by treating them per se as assets (Zur and Squires 2015). The analogy between nonpracticing entities and real estate agents, however, becomes tenuous, as the latter arguably do not litigate against property owners or file for invalidity of a title at the land registry. Although the assetization of legal interests in the case of patents can be compared to the securitization of debt obligations, investment trusts, and real estate, they seem to be qualitatively different. Patents embody contingent, speculative, and forward-looking potential, which is not always already realized and which has a temporal finitude of twenty years in theory. The financial assetization of patents amounts to a double speculation based on the abstractions of property and investment.

Attending to the distinct, internal characteristics of different knowledge practices allows drawing better distinctions between different modes of financialization and distinguish them from a commodity logic. Sherman and Powers (1994, 477) wrote that "both the very possibility of practice and the possibility of interaction between different fields are to be found in structures of knowledge and classification associated with each particular field." An insight into such "structures of knowledge and classification" requires both inside and outside perspectives. Whereas an interdisciplinary perspective is useful for tracing complex problematizations, attentiveness to the internal logics and the specific rhetorical and material practices of knowledge structures, such as law or finance, can identify the texture of composition and dynamic of a problematization (Kang 2018). Legal, financial, and scientific knowledges may intersect in issues relating to patents, but also they may not always do so. A cross-disciplinary patent scholar needs to be as attentive to the lack of interactions as suspect overarching claims of co-production. In this particular context, it may require the unpacking of hasty analogies to diagnose the specific ways in which patents act as assets. Although the rhetoric of patent office, economists and political economists has the effect of naturalizing patents as assets, albeit maybe for different reasons, it is important to bear in mind that patents do not always turn into assets or into financial assets. Financialization also cannot be conflated with assetization, for there are other kinds of asset values that are not necessarily financial. Patents and financialization do not always go hand in hand. Below I delineate some recent controversies where patents did not act as financial assets.

Patents have been objects of contention in the scholarship on bioeconomies (Birch 2017). Although patents in pharmaceutical and biotechnological products have been implicated in the development of bioeconomies, the financial assetization of patents needs to be distinguished from other modes of financialization in biotechnology companies. Recent scandals surrounding Turing, Valeant, and Mylan (Glabau 2017) have in common that they were owned by vested financial interests that demand high returns: they all have, or have had, private equity or hedge funds as main shareholders. One of the ways to increase return for these investors was by increasing the price of the product—by 5,000 percent in the case of Turing's Daraprim. But this price hike was independent of patent protection, which had already expired (*New York Times* 2015). Another common way to increase the return on investment is to heavily leverage the company. This was the case with Valeant, which was in threat of default and under SEC investigation (*Financial Times* 2016b).[11]

Patent validity shapes the pricing of patented drugs and company valuations, but so do a number of other factors, such as shareholder structure, debt-to-equity ratio, marketing, and distribution channels. Also, patents in this context are still operating as strategic business assets or as methods of cash generation, but are not necessarily primarily understood as *financial* instruments. Some of the most controversial recent price hikes were independent of patent protections, as in the cases of Mylan's EpiPen or Turing Pharmaceutical's Daraprim. EpiPen's chemical compound was not patented, but it was linked to a patented delivery device, which arguably could be uncoupled from the compound itself. Patents on a medicine might have long expired, but the marketing and distribution channels might be closely controlled and inaccessible so that a generic substitute would struggle to find distribution or that a small patient number would be seen as not worthwhile to produce a generic for (Sunder Rajan 2011; Peterson 2014). These observations do not negate the fact that patents are used as exclusionary monopoly forces to extract profits by sometimes exorbitant prices that cannot be paid by patients and health insurance systems while advancing unsubstantiated justifications referring to research and development costs (Love 2012). The overlap between products of knowledge, their monetization, and the use of legal rights for financial motives understandably causes unease and worry. But careful analysis is needed to differentiate between situations where patents are, or are not, the exclusive mechanisms by which profits in the biotech-pharmaceutical industry are realized.

Hedging Law

I have distinguished between an understanding of patents as a source of profits via commodification and patents as financial assets, the value of which does not necessarily derive from the exploitation of an invention, such as a drug compound, but rather from a speculation about the solidity of the legal property right itself. Treating patents as financial assets also entails a distinction between the value of the object of a property right (invention in the case of a patent) and the value of the property right (a patent).

Financial assetization of patents engages another level of abstraction on top of patent law's soft abstraction of inventive labor to an alienable commodity: law as the object of financial speculation. It is one thing to grant patent monopoly to a pharmaceutical manufacturer for the development of an antiviral drug. It is quite a different value proposition for patents to be implicated in a business model that encourages financial profits on the basis of speculating about the strength, demand, and enforceability of legal property rights and valued without much consideration of the original invention for which the patent is granted. In such a practice, the commodity to be traded and exchanged is not the invention; law itself becomes the asset and the commodity to go long or short on. The very operation of patent law becomes reconfigured as a matter of hedging for or against a patent's legal force.

The value of a patent as a financial asset is determined by hedging risk and time: projecting margins of stock price or company value fluctuations caused by a potential legal outcome. These hedges can consist in short-selling or going long on a company stock coupled with a challenge to an existing patent and estimating the level of damage awarded in litigation or a settlement in patent disputes. These considerations drive the threats of litigations of patent holders, both the business operations or nonpracticing entities who speculate on whether a patent right will be asserted by way of legal disciplinary power (e.g., a letter drawing on legal language or a threat of litigation), or through adjudication in the Patent Office, or by the courts. This has been apparent in the rise in applications for so called *inter partes* review at the USPTO, a process which was introduced by the America Invents Act in 2011. The *inter partes* review allows challenges to the validity of patents before the Patent and Trials Appeal Board (PTAB).

An example of the financial asset logic using the legal system as a market mechanism were the challenges against patents brought by the confusingly named Coalition for Affordable Drugs (CFAD), which consisted of hedge funds managed and owned by Kyle Bass, founder and principal of Hayman Capital Management LP. Bass brought thirty-six challenges to existing pharmaceutical patents. Bass argued that the purpose of the *inter partes* review applications was to make drug costs more affordable by invalidating unjustified patents, while at the same time stating that there was nothing wrong as such in raising patent validity challenges with a profit motive:

> The CFAD stated that "Celgene's motion [the holder of the challenged patent]… makes the curious argument that filing IPR petitions with a profit motive constitutes an 'abuse of process.' Yet at the heart of nearly every patent and nearly every IPR, the motivation is profit.…The U.S. economy is based largely on the notion that individual self-interest, properly directed, benefits society writ large." (CFAD's response, reported in Sidak and Skog, 2015, 124*ff.*)

The question of whether inventive knowledge should be seen as an origin from which profit and rent ought to be derived through layers of abstraction and intermediation is a normative one, and it cannot be neatly separated from the question of states of knowledge in financialized capitalist markets. Biotechnology has never been pure science (Thackray 1998). In *Biogen v. Medeva* (1996), one of the landmark cases in biotechnology patent disputes reaching the highest court in the UK, Lord Hoffman buried whatever remained of the belief in the separation between pure and applied sciences by stating that scientific progress need not be uncommercial in motives.[12] This juxtaposition, or a perceived alliance between science and commerce, has come a long way twenty years after *Biogen* and has morphed into a web of interests between technoscience and finance (Mirowski 2011).

Although Henry and Stiglitz (2010) have argued that challenging a patent is a "public good" because "there is an undersupply of public goods—implying that there will be too many patents granted because too few will be challenged" (Baker et al. 2017, 11), it is doubtful that shorting company shares by placing a bet on an adjudicatory outcome constitutes a "public good." Kyle Bass's financial vehicle, CFAD, contested the validity of thirty-six patents, eleven of which the USPTO dismissed outright as abuse of the institution. Seven applications for review were accepted, but they were not regarded as having significant financial value. By February 2016, Bass was reported to have returned most of the $700 million that he raised for short-selling pharmaceutical stocks but maintained that "we have all the capital that we need to pursue everything to its logical conclusion at the patent office.…we are not stopping" (*Financial Times* 2016a). Bass argued that even if he was to short-sell patent holders' shares and make profit from lower share prices, it would not be an abuse of process but rather contribute to market efficiency, referring to the reasoning of the Securities and Exchange Commission, which had stated that "short sellers who short companies with overvalued stock can actually add to stock pricing efficiency by informing the market of the true economic value of those companies" (Sidak and Skog 2015, 125). The belief that the "true economic value" of the patents is reproduced in company share prices is reflected in index-tracking investment funds, such as the ones offered by Ocean Tomo. For example, the Ocean Tomo 300 Patent Index tracks the value of patents in a portfolio of three hundred companies that are deemed to hold the most valuable patents relative to their book value. It is said to have outperformed the S&P 500 from its inception up to 2015 by 1,620 basis points, which Ocean Tomo attributes to its emphasis put on patent valuation through "its proprietary Intellectual Property valuation methodologies," according to its investments brochure.

Patents act as financial assets by linking financial hedges (measured by expected share price fluctuations) to legal decisions and outcomes. This practice is significantly different in its logic from other quantitative modelings of patent value. Here, value is enacted through a legal process of adjudication and not only through discounted cashflow valuation or option pricing models. The temporality in which value is seen to reside is also different. Valuing patents as real options is based on the premise that they are potentially valuable assets to possess in the future. Bass's patent validity challenges, in contrast, function by betting against the legality of past legal processes: they are speculations on the administrative and procedural strengths and weaknesses of the patent law system itself. Effectively this means that the patent law process—starting from the patent application and review process, to the robustness of the PTAB, the inclination of the Federal Circuit and the judiciary—is conceived and framed as a market in which financial value can be created. The legal form of a property right is unpacked and challenged as proxies of share prices. Patents are seen as proxy instruments for market-arbitrage. If one takes the SEC's claim of "true economic value" as an analogy, the question is whether these hedges will indeed make the patent system more efficient or whether such financial arbitraging of legal processes is incompatible with the functioning of the legal system itself, the rationality of which is based on norms than probabilities. Another way of reading this development could be that the free market premise on which modern patent law has been based has been realized to its full extent.

Conclusion: Patent Law as Market Subject and Object

Lazzarato (2014, 46) argues that the meaning of "intellectual" in capitalism has shifted from denoting a human mental capacity to the "'intellectual' and physical performance of machines, protocols, organization, software or systems of signs, science, and so on." Whatever remains of the attribute "intellectual" in intellectual property in its transformation into a financial asset refers to itself: law as an assemblage of very specific knowledge techniques and practices. Going beyond the realization that assetization occurs, further analyses are needed of the concrete processes of assetization and their implications. The turning of patents into assets is not only a symptom of capitalization and assetization of technoscience with novel ways and degrees of speculative financialization. The assetization of patents represents a new frontier, a novel financial "innovation," affecting a knowledge practice that hitherto had not been regarded as an object of speculation: law. Speculation about legal outcomes and decisions drives this novel financial asset.

The financial assetization of patents entails different modes of rationality and temporality than those of commodification. The rationality that drives patents' financialization is probability (as opposed to monopoly), and its temporal mode is one of speculation (rather than of a recoupment, reward, or promise). The value of the patents as financial assets is neither measured by primary reference to the inventive thing itself nor by its surplus value as a commodity. Rather the value of patents as assets rests on the legal fiction of intellectual property and specific valuation practices, which vary in their techniques and settings, for example, as real options or by modeling short-sell hedges. The financial assetization of patents is an apt example of what Joseph Vogl (2010, 80) has called "capital's credo," the institution of a self-referential system in which the main referent is no longer a tangible good but an intangible mode of legal credit in the form of intellectual property right. Here the flow of exchange is no longer "good—legal credit—good," but "money credit—legal good—money credit."

The main referent in this transaction flow is neither the good of a commodity nor its original referent, an invention, but financial capital in search for the next yield. Free of tangible referents, illusions of value circulate and are effectively transformed into determinants of economic relations within the narratives of financial markets. To the double abstraction in patent law's operation—conjuring an abstract legal right out of something which is intangible, the inventive essence—another layer of abstraction is added: the one of financial assetization. These triple layers of abstraction reveal the legal system in its full contingency and its complex entanglements with the financialized economy. What is ultimately at stake is the financialization of the legal system itself. Legal forms have turned into financial assets. Whereas law institutes financial capitalism (Pistor 2014), here we see law being financialized itself. Financial capitalism has turned on its own constitutive foundation, law. Whereas law has enabled the creation of a market in patents as assets, but it has now become a financialized market object itself.

Acknowledgments

I am indebted to Kathy Bowrey for her close and critical reading of an earlier version of this chapter, as well as the anonymous reviewers for their constructive comments.

Notes

1. See Sunder Rajan (2011) for the constitutionalization of intellectual property issues in the Indian Supreme Court and an overview of the different developments and regulations of the major patent law systems. Khan (2008) notes the German patent system's historical prohibition on the patenting of food, chemical, or pharmaceutical products.

2. Thanks to Kathy Bowrey for pointing out this underlying premise so clearly.

3. Modern patent law in the form of registration and examination as we know it has only existed arguably since the introduction of the Patent Act of 1836 in the US. The 1836 Act introduced the requirements of written specification, a predominantly text-based understanding of novelty and prior art reflecting the belief that there was a bargain to be struck between the monopoly right of a patent and the public interest to know about inventions through their disclosure in patent documents (Biagioli 2006).

4. 569 U.S. 576 (2013), https://www.supremecourt.gov/opinions/12pdf/12-398_1b7d.pdf.

5. This distinction often becomes muddled when patents are equated with inventions, as they are not the same. Patents and inventions are different entities: the former is a limited monopoly right, an intangible legal form, and the other is its object.

6. The Ocean Tomo 300 ® Patent Index (OT3000), http://www.oceantomo.com/ocean-tomo-300/. Ocean Tomo also runs an IP auction service, in which patents are auctioned online.

7. There is no clear dichotomy and alignment of interests between practicing companies that invent and license, on the one hand, and nonpracticing entities, who act like IP brokers or investors, managing IP as assets, on the other. Even university spin-offs, such as Oxford Sciences Innovation in the UK, act as IP management service, investment fund and incubator. Patent brokers such as IV and RPX, do not only act as intermediaries but also as principal investors in business with patents or as incubators. Straddling the private-public divide, France Brevet, is a government-backed IP broker and fund, reflecting a state-backed industrial policy through IP, as does Singapore IP Office, which stands in as the ultimate security for loans given for IP assets as collaterals. Other companies, such as Intellectual Ventures (backed by Google) have the business model of defensively purchasing patents in order to prevent NPEs from doing so and to challenge other NPE claims through *inter partes* review at the US Patents and Trademarks Office (USPTO).

8. For a general patent valuation overview, see Kamiyama et al. (2006) and PricewaterhouseCoopers (2007); for econometric use of patent renewal data for overall macroeconomic value of patents, see, for example, Pakes (1986); for option pricing model for individual patent as asset valuation, see Damodaran (2006, chapter 12) and Gu and Lev (2008).

9. For example, in Singapore, patents have been accepted as a collateral for cash loans from banks, but seem to be part of an industrial policy which is backed up by the government. See the press release from the Intellectual Property Office of Singapore (IPOS), https://www.ipos.gov.sg/media-events/press-releases/ViewDetails/cash-for-intellectual-property-through-loan-financing-now-a-reality-in-singapore/. Patents can be theoretically accepted as collaterals from banks, but this is rare due to the speculative nature of their valuation.

10. Well-known players offering patent funds are IV, RPX, and Ocean Tomo. Deutsche Bank and Credit Suisse had launched patent portfolio funds in 2008. The Deutsche one raised a total of around €300 million but filed for bankruptcy in 2010. Their short-livedness indicates that the assetization of patents is not entirely abstract and purely financial as in the case of REITs due to pricing intransparency in license transactions and the difficulty in comparing patent values.

11. Valeant had a junk bond credit rating in 2015. In 2016, it was recommended to file for bankruptcy. By March 2017, its shares had lost 95 percent in value since 2015. Valeant now still has a debt to equity ratio of 12/88.

12. [1996] UKHL 1, http://www.bailii.org/uk/cases/UKHL/1996/18.html.

References

Aragon, L., and Leach, J. 2008. Arts and Owners: Intellectual Property Law and the Politics of Scale in Indonesian Art. *American Ethnologist* 35 (4): 607–631.

Baker, D., Jayadev, A., and Stiglitz, J. 2017. *Innovation, Intellectual Property, and Development.* AccessIBSA: Innovation and Access to Medicines in India, Brazil and South Africa, http://ip-unit.org/wp-content/uploads/2017/07/IP-for-21st-Century-EN.pdf.

Barnett, J. 2016. *From Patent Thickets to Patent Networks: The Legal Infrastructure of the Digital Economy*. Working Paper 121: University of Southern California Legal Studies Working Paper Series.

Bellido and Kang. 2016.

Bessen, J., and Meurer, M. 2008. *Patent Failure: How Judges, Bureaucrats, and Lawyers Put Innovators at Risk*. Princeton, NJ: Princeton University Press.

Bessen, J., and Meurer, M. 2014. The Direct Costs from NPE Disputes. *Cornell Law Review* 99 (2): 387–424.

Bhandar, B. 2017. Registering Interests: Modern Methods of Valuing Labour, Land, and Life. In *Searching for Contemporary Legal Thought*, edited by J. Desaultes-Stein and C. Tomlins, 290–311. Cambridge: Cambridge University Press.

Biagioli, M. 2006. Patent Republic: Representing Inventions, Constructing Rights and Authors. *Social Research* 73 (4): 1129–1172.

Biagioli, M. 2009. Nature and the Commons: The Vegetable Roots of Intellectual Property. In *Living Properties: Making Knowledge and Controlling Ownership in the History of Biology*, edited by D. Kevles and J.-P. Gaudilliere, 241–250. Berlin: Max Planck Institute for the History of Science.

Biagioli, M. 2014. Plagiarism, Kinship and Slavery. *Theory Culture Society* 31 (2/3): 65–91.

Bilir, L. K., Moser, P., and Talis, I. 2011. Do Patent Treaties Encourage Technology Transfer? Evidence from the Paris Convention, http://ssrn.com/abstract=1893052.

Birch, K. 2017. Rethinking Value in the Bio-economy: Finance, Assetization, and the Management of Value. *Science, Technology, and Human Values* 42 (3): 460–490.

Boltanski, L., and Thévenot, L. 2006. *On Justification: Economies of Worth*. Princeton, NJ: Princeton University Press.

Boyle, J. 2010. *The Public Domain*. New Haven, CT: Yale University Press.

Bracha, O. 2005. *Owning Ideas: A History of Anglo-American Intellectual Property*. SJD Thesis. Cambridge, MA: Harvard Law School.

Burk, D., and Lemley, M. 2009. *The Patent Crisis and How the Courts Can Solve It*. Chicago: University of Chicago Press.

Cassier, M. 2005. Appropriation and Commercialization of the Pasteur Anthrax Vaccine. *Studies in History and Philosophy of Science Part C* 36 (4): 722–742.

Cassier, M., and Correa, M. 2014. Access to Medicines in Developing Countries: Ethical Demands and Moral Economy. *Developing World Bioethics* 14: ii–viii.

CBC News. 2018. Drug Patents Could Be Canada's Special Weapon in U.S. Trade Dispute, June 16, 2018, https://www.cbc.ca/news/health/second-opinion-trade-dispute-canada-us-drug-patents-intellectual-property-1.4708630.

Coombe, R. 1998. *The Cultural Life of Intellectual Properties*. Durham, NC: Duke University Press.

Damodaran, A. 2006. *Damodaran on Valuation*. New York: Wiley.

Dasgupta, P., and Stiglitz, J. 1988. Potential Competition, Actual Competition, and Economic Welfare. *European Economic Review* 32: 569–577.

Drahos, P., and Braithwaite, J. 2002. *Information Feudalism: Who Owns the Knowledge Economy?* London: Earthscan.

Dreyfuss, R., and Frankel, S. 2015. From Incentive to Commodity to Asset: How International Law Is Reconceptualizing Intellectual Property. *Michigan Journal of International Law* 36 (4): 557–602.

Drucker, P. 1969. *The Age of Discontinuity*. New York: Harper and Row.

Edelman, B. 1979. *Ownership of Image: Elements of a Marxist Theory of Law*. London: Routledge Kegan Paul.

Escoffier, L., and Kasznik, E. 2012. The Use of IP valuation in IP transactions: A Global Survey of IP Brokers. *European IPR Helpdesk Bulletin* (January-March, No. 4): 4–5.

European Commission, PatVal-EU Project. 2005. *The Value of European Patents. Evidence from a Survey of European Inventors, Final Report,* http://ec.europa.eu/invest-in-research/pdf/download_en/patval_mainreportandannexes.pdf.

European Patent Office. 2017. *Patent Information News 2017,* http://www.epo.org/service-support/publications.html?pubid=132#tab3.

Financial Times. 2015. Kyle Bass Plans Legal Action on Pharma Patents. *Financial Times,* January 7, 2015.

Financial Times. 2016a. Kyle Bass Returns Funds Amid Retreat on Pharma Shorting Campaign. *Financial Times,* February 23, 2016.

Financial Times. 2016b. Valeant: The Harder They Fall. *Financial Times,* March 28, 2016.

Financial Times. 2017. Big Tech Makes Vast Gains at Our Expense. Comment by Rana Foroohar. *Financial Times,* September 17, 2017.

Fisk, C. 2009. *Working Knowledge: Employee Innovation and the Rise of Corporate Intellectual Property, 1800–1930*. Chapel Hill, NC: University of North Carolina Press.

Forbes. 2016. Why Did Mylan Hike EpiPen Prices 400%? Because They Could. *Forbes,* August 21, 2016.

Fortun, M. 2008. *Promising Genomics: Iceland and deCode Genetics in a World of Speculation*. Berkeley, CA: University of California Press.

Galvez-Behar, G. 2016. Louis Pasteur, entrepreneur. Pour une histoire économique des mondes savants, https://halshs.archives-ouvertes.fr/halshs-01267638v4/document.

Gaudilliere, J.-P., and Loewy, I., eds. 1998. *The Invisible Industrialist: Manufacture and the Construction of Scientific Knowledge.* London: Palgrave Macmillan.

Glabau, D. 2017. Conflicting Assumptions: The Meaning of Price in the Pharmaceutical Economy. *Science as Culture* 26 (4): 455–467.

Griliches Z. 1990. Patent Statistics as Economic Indicators: A Survey. *Journal of Economic Literature* 28: 1661–1707.

Griliches, Z., Pakes, A., and Hall, B. 1987. The Value of Patents as Indicators of Inventive Activity. In *Economic Policy and Technological Performance,* edited by P. Dasgupta and P. Stoneman, 97–124. Cambridge: Cambridge University Press.

Gu, F., and Lev, B. 2008. Markets in Intangibles: Patent Licensing. *NYU Working Paper* No. 2451/27465.

Hagiu, A., and Yoffie, D. 2013. The New Patent Intermediaries: Platforms, Defensive Aggregators, and Super Aggregators. *Journal of Economic Perspectives* 27 (1): 45–66.

Haskel, J., and Westlake, S. 2018. *Capitalism without Capital: The Rise of the Intangible Economy.* Princeton, NJ: Princeton University Press.

Hayden, C. 2003. *When Nature Goes Public.* Princeton, NJ: Princeton University Press.

Heller, M., and Eisenberg, R. 1998. Can Patents Deter Innovation? The Anticommons in Biomedical Research. *Science* 280 (5364): 698–701.

Henry, C., and Stiglitz, J. 2010. Intellectual Property, Dissemination of Innovation, and Sustainable Development. *Global Policy* 1 (1): 237–251.

Homburg, E., Travis, A. S., and Schroeter, H. G. 1998. *The Chemical Industry in Europe, 1850–1914.* Dordrecht: Springer.

Intellectual Asset Management Magazine. 2008. Mixed View on German Patent Investment Funds. *Intellectual Asset Management Magazine,* March 28, 2008.

Israel, P. 2000. *Edison: A Life of Invention.* New York: Wiley.

Jaffe, A. 1989. Real Effects of Academic Research. *American Economic Review* 79 (5): 957–970.

Jaffe, A., and Lerner, J. 2006. *Innovation and Its Discontents: How Our Broken Patent System is Endangering Innovation and Progress, and What to Do about It.* Princeton. NJ: Princeton University Press.

Kamiyama, S., Sheehan, J., and Martinez, C. 2006. Valuation and Exploitation of Intellectual Property. *OECD Science, Technology and Industry Working Papers,* No. 2006/05. Paris: OECD.

Kang, H. Y. 2012. Science Inside Law: The Making of a New Patent Class in the International Patent Classification. *Science in Context* 25: 551–594.

Kang, H. Y. 2015. Patent as Credit: When Intellectual Property Becomes Speculative. *Radical Philosophy* 194: 29–37.

Kang, H. Y. 2018. Law's Materiality: Between Concrete Matters and Abstract Forms, or How Matter Becomes Material. In *Routledge Handbook for Law and Theory*, edited by A. Philippopoulos-Mihalopoulos, 453–474. Abingdon: Routledge.

Kang, H. Y. 2019. Ghosts of Inventions: Patent Law's Digital Mediations. *History of Science. Special Issue: Technologies of the Law/Law as a Technology* 57 (1): 38–61.

Kang, H. Y., and Bellido, J. 2016. In Search of a Trade Mark: Search Practices and Bureaucratic Poetics. *Griffiths Law Review* 25 (2): 147–171.

Kapczynski, A., Park, C., and Sampat, B. 2012. Polymorphs and Prodrugs and Salts (Oh My!) *PLoS ONE* 7 (12): e49470.

Kelty, C. 2008. *Two Bits: The Cultural Significance of Software*. Durham, NC: Duke University Press.

Khan, Z. B. 2008. An Economic History of Patent Institutions. *EH.Net Encyclopedia*, http://eh.net/encyclopedia/an-economic-history-of-patent-institutions/.

Kostylo, J. 2008. Commentary on Johannes of Speyer's Venetian Monopoly (1469). *Primary Sources on Copyright (1450–1900)*, www.copyrighthistory.org.

Krikorian, G., and Kapczynski, A. 2010. *Access to Knowledge in the Age of Intellectual Property*. Cambridge, MA: MIT Press.

Lanjouw, J. O., Pakes, A., and Putnam, J. 1998. How to Count Patents and Value Intellectual Property: Uses of Patent Renewal and Application Data. *Journal of Industrial Economics, XLVI*(4): 405–433.

Lanjouw, J. O., and Schankerman, M. 1997. Stylised Facts of Patent Litigation: Value, Scope and Ownership. *NBER Working Paper Series*, Working Paper No. 6297.

Lanjouw, J. O., and Schankerman, M. 2001. Characteristics of Patent Litigation: A Window on Competition. *RAND Journal of Economics* 32 (1): 129–151.

Latour, B. 1993. *We Have Never Been Modern*. Cambridge, MA: Harvard University Press.

Lazzarato, M. 2014. *Signs and Machines: Capitalism and the Production of Subjectivity*. Cambridge, MA: MIT Press.

Leydesdorff, L. 2004. The University-Industry Knowledge Relationship: Analyzing Patents and the Science Base of Technologies. *Journal of the American Society for Information Science and Technology* 55 (11): 991–1001.

Love, J. 2012. *The De-linkage of R&D Costs and Drug Prices through the Prize Fund for HIV/AIDS.* Testimony at Hearing before the United States Senate, Committee on Health, Education, Labor and Pensions, Subcommittee on Primary Health and Agency on the High Cost of High Prices for HIV/AIDS Drugs and the Prize Fund Alternative, May 15, 2012, https://www.keionline.org/sites/default/files/testimony_JamesLove_15May2012_prizes_HELP_subject2revision.pdf.

MacKenzie, D. 2006. *An Engine, Not a Camera: How Financial Models Shape Markets.* Cambridge, MA: MIT Press.

Menell, P. 2018. Economic Analysis of Intellectual Property Notice and Disclosure. In *Research Handbook on the Economics of Intellectual Property Law: Vol. I, Theory,* edited by B. Depoorter and P. Menell. Cheltenham: Edward Elgar.

Mirowski, P. 2011. *ScienceMart.* Cambridge, MA: Harvard University Press.

Moser, P. 2005. How Do Patent Laws Influence Innovation? Evidence from Nineteenth-Century World Fairs. *NBER Working Paper Series,* Working Paper No. 9909.

Myers, G. 1995. From Discovery to Invention: The Writing and Rewriting of Two Patents. *Social Studies of Science* 25 (1): 57–105.

Muniesa, F. 2014. *The Provoked Economy.* London: Routledge.

Muniesa, F., Doganova, L., Ortiz, H., Pina-Stranger, A., Paterson, F., Bourgoin, A., Ehrenstein, V., Juven, P.-A., Pontille, D., Saraç-Lesavre, B., and Yon, G. 2017. *Capitalization: A Cultural Guide.* Paris: Presses des Mines.

Nelkin, D. 1984. *Science as Intellectual Property.* New York: Macmillan.

New York Times. 2012. Kodak to Sell Digital Imaging Patents for $525 Million. *New York Times,* December 19, 2012.

New York Times. 2015. Drug Goes from $13.50 a Tablet to $750 Overnight. *New York Times,* September 20, 2015.

New York Times. 2016. Mylan Raised EpiPen's Price Before the Expected Arrival of a Generic. *New York Times,* August 24, 2016.

OECD. 2009. *Patent Statistics Manual.* Paris: OECD.

Pakes, A. 1986. Patents as Options: Some Estimates of the Value of Holding European Patent Stocks. *Econometrics* 54 (4): 755–784.

Peterson, K. 2014. *Speculative Markets: Drug Circuits and Derivative Life in Nigeria.* Durham, NC: Duke University Press.

Pistor, K. 2013. A Legal Theory of Finance. *Journal of Comparative Economics* 41 (2): 315–330.

Pottage, A., and Sherman, B. 2010. *Figures of Invention: A History of Modern Patent Law.* Oxford: Oxford University Press.

Power, M. 1992. The Politics of Brand Accounting in the United Kingdom. *European Accounting Review* 1: 39–68.

PricewaterhouseCoopers. 2007. *Technology Executive Connections: Exploiting Intellectual Property in A Complex World*, https://www.wipo.int/ip-outreach/en/tools/practice/details.jsp?id=865.

Pricewaterhouse Coopers, Zareer Pavri. 1999. *Valuation of IP Assets. The Foundations for Risk Management and Financing*, http://www.bvstrategy.com/Intell3_99.pdf.

Raunig, G. 2013. *Factories of Knowledge, Industries of Creativity*. Cambridge, MA: semiotext(e)/Intervention, MIT Press.

Ricketson, S. 2015. *The Paris Convention for the Protection of Industrial Property: A Commentary*. Oxford: Oxford University Press.

Risch, M. 2013. Patent Portfolio as Securities. *Duke Law Journal* 63: 89–154.

Ryan, M. 1998. *Knowledge Diplomacy: Global Competition and the Politics of Intellectual Property*. Washington, DC: Brookings Institution.

Sherman, B., and Powers, M. 1994. Law, Accounting and the Emergent Positivity of Intangible Property. *Social and Legal Studies* 3: 477–495.

Sidak, J. G., and Skog, J. O. 2015. Attack of the Shorting Bass: Does the *Inter Partes* Review Process Enable Petitioners to Earn Abnormal Returns? *UCLA Law Review Discourse* 63: 120–155.

Steen, K. 2000. German Chemicals and American Politics, 1919–1921. In *The German Chemical Industry in the Twentieth Century*, edited by J. Lesch, 323–346. Dordrecht: Kluwer.

Strathern, M. 1999. *Property, Substance and Effect: Anthropological Essays on Persons and Things*. London: Athlone.

Sunder Rajan, K. 2006. *Biocapital*. Durham, NC: Duke University Press.

Sunder Rajan, K. 2011. Property, Rights, and the Constitution of Contemporary Indian Biomedicine: Notes from the Gleevec Case. *Social Research* 78 (3): 975–998.

Thackray, A., ed. 1998. *Private Science: Biotechnology and the Rise of Molecular Sciences*. Philadelphia: University of Pennsylvania Press.

Trajtenberg, M. 2002. *Patents, Citations, Innovations*. Cambridge, MA: MIT Press.

Tucker, C. 2014. The Effect of Patent Litigation and Patent Assertion Entities on Entrepreneurial Activity, https://ssrn.com/abstract=2457611.

UK Department for Business, Innovation and Skills. 2016. *Success as a Knowledge Economy: Teaching Excellence, Social Mobility and Student Choice*, https://assets.publishing.service.gov.uk/governmet/uploads/system/uploads/attachment_data/file/523546/bis-16-265-success-as-a-knowledge-economy-web.pdf.

UK Intellectual Property Office. 2017. *Promoting Innovation and Growth. The Intellectual Property Office at Work 2016–2017*, https://assets.publishing.service.gov.uk/government/uploads/system/uploads/attachment_data/file/653493/Innovation-and-growth-report-2016-17.pdf.

Vogl, J. 2014. *Specter of Capital*. Stanford, CA: Stanford University Press.

Wagner, R. P., and Parchomovsky, G. 2005. Patent Portfolios. *University of Pennsylvania. Law Review* 154: 1–77.

Zur, E., and Squires, J. 2015. Why Investment-Friendly Patents Could Spell Trouble for Trolls, http://knowledge.wharton.upenn.edu/article/why-investment-friendly-patents-spell-trouble-for-trolls/.

3 Datassets: Assetizing and Marketizing Personal Data

Thomas Beauvisage and Kevin Mellet

Introduction

In a 2011 report entitled "Personal Data: The Emergence of a New Asset Class," the World Economic Forum (WEF) stated that "personal data is generating a new wave of opportunity for economic and societal value creation.... As some put it, personal data will be the new 'oil'—a valuable resource of the 21st century. It will emerge as a new asset class touching all aspects of society" (Schwab et al. 2011, 5). Starting from this observation, the report called for the establishment of a new data infrastructure, which would simultaneously give individuals high control over their personal data and facilitate the flow and exchange of data so far retained in isolated silos among individuals, companies, and public institutions. The report predicted that personal data would constitute a new form of currency for individuals—one that they would be free to manage themselves: "In practical terms, a person's data would be equivalent to their 'money.' It would reside in an account where it would be controlled, managed, exchanged and accounted for just like personal banking services operate today" (Schwab et al. 2011, 10).

Admittedly, the economic importance of personal data is not new. In the 1970s (to stick to contemporary times), private players developed database management software, and some of them, which later became known as data brokers, established the basis for the trading of lists (of addresses, contacts, profiles, credit reports, etc.). It was during this period that an academic literature on data and privacy (Posner 1981) and modern legislation on the protection of personal data developed (Fair Credit Reporting Act, USA 1970; Loi Informatique et Libertés, France 1978). That said, the recent and combined accelerations in the areas of networks, storage, and computing, and the multiplication of sensors that keep a record of various traces have led to a renewed interest in data, sometimes labeled as the "big data revolution" (Kitchin 2014). As pointed out by Schneier, "in the normal course of their operations, computers continuously document what they're doing" (2015, 15). Big data systematizes the apprehension of data and represents a valuable resource requiring attention and investment.

The big data revolution raises a host of questions. How to make money, and more generally extract value, from personal data? How concretely are personal data transformed into economic assets that generate revenue streams and value? How does this new asset class emerge? The WEF provides a typically neoliberal answer to this question, but it is not the only possible one. Arguing for a supposed "exceptionalism" of big data (Nissenbaum 2017), large technology companies and a multitude of start-ups collect huge amounts of user data and put together teams of data scientists to analyze and extract value from these data. Traditional companies have gone through an assetization movement in recent years as well. They have set up data management entities headed by chief data officers and assigned them the task of identifying valuable first-party data within their complex information systems in order to derive revenue streams from them.

Building on a decade-long inquiry into the online marketing and advertising industry, this chapter analyzes the process by which consumer data have come to be considered as valuable, and sometimes tradable, assets. In the introductory chapter, Birch and Muniesa defined assets as "something that can be owned or controlled, traded, and capitalized as a revenue stream, often involving the valuation of discounted future earnings in the present." This precise characterization of assets appears as a distant horizon when one is interested in personal data. Indeed, the technical, legal, and economic contours of personal data appear hazy and fluctuating. Moreover, contrary to what is observed in finance, there is no set of standards and instruments for measuring, assessing, and discounting future revenue streams from personal data. Finally, market-based trading of personal data by data brokers and similar actors is only a small part of a much larger transformation that implies the production, activation, and flow of data, at a large scale. As underlined by Muniesa et al., anthropology and sociology typically focus on "commercial valuation—things being valued as they are bought and sold, and hence valued as commodities in the market" (Muniesa et al. 2017, 13), but pay little attention to the process of capitalization.

We argue that the ability to capitalize personal data in the present is the result of a versatile and uncertain process of assetization. This chapter seeks to unpack this process and to identify and qualify various data activation regimes and the resulting statuses of data as assets, or "datassets" as we name them.

The online marketing and advertising industries are a striking example of the dynamics of data assetization, a movement also observed in areas such as healthcare (Ebeling 2016), insurance (McFall 2014), and government assistance (Eubanks 2018). They provide a notable illustration of something that economic institutions and mainstream economic thought have considered explicitly as an asset, while letting firms and market intermediaries perform this claim. Our examination of how economic players have succeeded or failed in their attempts to assetize personal data unpacks the temporal and entrepreneurial dynamics of assetization, their material and discursive devices, and highlights the plurality of asset forms that emerge from this dynamic.

The chapter is structured as follows. The first section presents a brief survey of the literature. Next, we analyze the repeated and unsuccessful attempts to create a consumer-to-business (C2B) market for personal data, and the underlying conception of personal data marketization as a trade-off between privacy and the benefits of tailored services. This failure can be considered as a performativity failure of economic models and related experiments. In the third section, we investigate the actual markets for personal data, a business-to-business (B2B) activity strongly linked with marketing and advertising services. We show that personal data can be commodified and traded, especially by data brokers, but as resources adding value to advertising products in the form of contacts, segments, or attributes. We finally argue that the spreading of tracking technologies and data management platforms (DMPs) inside private companies contributes to transform datasets into "datassets," that can either compete or articulate with third-party, commodified data.

Putting Personal Data on the Research Agenda

Within the field of marketing and advertising there is an emerging literature seeking to open the black box of data marketing by describing its intermediaries, products, processes, and organizing principles. In his book

The Daily You, Turow (2012) investigates the development of the online advertising market. He presents an entire industry engaged in a process of splitting Internet users into "targets and waste." Turow argues that a process of personalization is under way, based on the creation of extremely detailed databases on the characteristics and tastes of Internet users. At the end of his book, he focuses on data brokers, a set of emerging actors specialized in the collection, aggregation, processing, and selling of personal data, which tend to gain importance in the ecosystem of online advertising. Data brokers, like Acxiom or Experian, are shrouded in secrecy: their activities were revealed to the public by an investigation conducted by the Federal Trade Commission (2014). The corresponding FTC report provides a valuable description of this industry: its data acquisition methods, its products (segments and scores for marketing, risk mitigation for finance, people search), and its customers. This report serves as a primary source for works that seek to account for the breadth and variety of forms of tracking and surveillance put in place by private companies in the 2010s (Christl and Spiekermann 2016). Conversely, some authors notice how increasing concern about privacy has been seized as a business opportunity by entrepreneurs (Milyaeva and Neyland 2016).

Other research examines the forms of stranglehold and appropriation at work on the part of this industry and questions the moral and legal legitimacy of this foreclosure. Ebeling (2016) is interested in how ordinary life events become objects of capitalization by data brokers. In an ethnographic inquiry into her "marketing baby," she describes the logics of capture implemented by data brokers and how they are coupled with a form of appropriation. For her, the transformation of private data into assets is based on an ownership claim. "Brokers argue that by 'adding value' to data through analyzing it, processing it, de-identifying it, and creating new instruments or data products out of 'raw' data, brokers firm up their ownership claims. A common refrain in the industry is that data are the 'new oil' and brokers are the 'processors' that refine it and make it into 'products'" (2016, 44).

Nissenbaum (2017) is interested in the legal arguments put forward by such big data promoters, especially those engaged in an ideological and political battle to eliminate the regulations covering the collection of data. According to them, it is the use of data, not its collection, that must be regulated, a stance Nissenbaum labels "big data exceptionalism." In other words, personal data constitute a store of (economic, social, etc.) future value, but the uncertainty about the nature and the modalities of extraction of this value are framed such that the recording of traces and constitution of giant databases should not be hindered. Following a complementary perspective, Fourcade and Healy (2017) point out that organizations follow an institutional data imperative to collect as much data as possible, which results in the constitution or reinforcement of a new form of übercapital built from these data, in particular in the form of rankings and individual scores.

These works provide a solid basis for analyzing how personal data are constructed as economic assets in industrial and market processes, although they do not focus on the devices put in place by corporations to transform data into assets. Rather, capitalization is often reduced to a binary sequence of operations of capture and marketization, and carried out by data brokers and other market intermediaries only.

Attempts to Marketize Personal Data

Around 2010, personal data appeared to be a new asset for a large range of economic players, strongly linked with the digital economy and often described as the "new oil" in a future fully digitized world. But the potential value of this new resource was still to be determined. To extract a share of this new value, a part of the digital economy considered personal data as a commodity, a standard elementary tradable good like oil, close to a currency. Such perspectives on personal data as tradable assets was reinforced by the forecasts of neoliberal institutions such as the World Economic Forum, which imagined a direct conversion of personal data into revenue streams, as we mentioned above.

"New Assets" as Possible Products and Markets in the Future

The WEF's initiative is not isolated. It took place in a doubly favorable context. On the one hand, in the digital economy, entrepreneurs glimpsed an opportunity to build business models around the marketing by individuals of their own data. On the other hand, within the civil society, individual rights groups such as the Electronic Frontier Foundation have supported initiatives aimed at empowering individuals against the gluttony of large digital service corporations. In this context, it seemed that there was a genuine prospect of a market for personal data in which Internet users would make money from selling information about themselves. In order to sketch the outlines of such a system, academic research and consumer surveys led by private companies attempted to appraise the market worth of personal data. These economic experiments all take the same approach, immersing Internet users in fictitious situations in which they have to weigh up the cost of disclosing information about themselves (age, income, Internet history) against a financial benefit (discount or revenue). The value of data is thus defined as the price—or its equivalent in service terms—at which individuals would agree to transfer specific pieces of private information. The aim of these initial valuations was to identify the data whose protection individuals valued most highly, as well as the third parties to whom they would consider selling them and at what price. It was also a way to assess the cost of privacy (see Table 3.1).

Table 3.1
Summary of studies on the worth of personal data item for individuals

Source	Average valuations
Huberman et al. (2005)	Age = \$57 Weight = \$74
Danezis et al. (2005)	Location ~ £27
Beresford et al. (2012)	Favorite color and year of birth = €1
Carrascal et al. (2013)	"Offline" information (age, address, economic status) ~ €25 Browsing history = €7 Interactions on social networks = €12 Search history = €2 Shopping = €5
Staiano et al. (2014)	Geolocation = €17 to €588 Communications = €3.40 to 51
Havas Média (2014)[a]	30% sell in the top band, i.e., "More than €500"
Orange (2014)[b]	Each piece of information ~ €15 (name, mobile number, children's ages, income, purchase history, contacts, etc.) Average total value = €170

[a]Havas Media. 2014. Les Français et leurs données personnelles, quelle place pour les marques? Corporate report.
[b]Orange. 2014. The Future of Digital Trust. A European Study on the Nature of Consumer Trust and Personal Data. Corporate report, http://www.orange.com/content/download/25973/582245/version/2/file/Report+-+My+Data+Value+-+Orange+Future+of+Digital+Trust+-+FINAL.pdf.

The study conducted by Carrascal et al. (2013) is a good example. The experiment was based on a plug-in developed for web browsers, which, in real time, asked a panel of participants (n=168) at what price they would be willing to sell information about their behavior and practices on the web. "Offline" information (age, address, economic status) had a high price, at around €25. The average price asked by participants for a permanent transfer of their browsing history, meanwhile, was €7, with variations depending on the information (€12 for interactions on social networks; €15.50 for visits to financial sites; €2 for search history; €5 for shopping).

Unsuccessful Attempts to Create a C2B Market for Personal Data

Some start-up companies explored this path, banking on the idea that digital players would respect the need to seek users' consent to provide their data and on the notion of attaching proprietary rights to personal data, thus turning it into an asset. We have identified five such players, in the United States (Personal, Datacoup, Personal Black Box), the United Kingdom (Handshake), and France (Yes Profile). These five companies were established between 2009 and 2013.

These five start-ups all used the same pitch, which can be summed up as follows: the business model of Internet giants is based on the unfair seizure of private information from which they make enormous profits. Instead, these start-ups would act as personal information brokers on behalf of their subscribers toward the advertising industry, redirecting profits and making users' consent a central tenet. The model adopted by these companies involved asking Internet users to entrust them with data from social networks, as well as entering a considerable amount of additional information about themselves, particularly about their purchase intents, or even taking part in focus groups and testing products. The potential gains these start-ups promised to web users were in some cases highly exaggerated: Handshake suggested the revenue stream could reach £15,000 (€19,000) per participant per year; Yes Profile estimated potential monthly income at €60 to €100, giving an annual total of up to €1,200; and Personal Black Box gave a range from $50 to $500 (€45 and €450). Datacoup, meanwhile, was more cautious, with its site advancing the figure of €8 a month, or around a hundred euros a year.

This business model never took off. Ultimately, it came up against one major stumbling block: the capture and sale of web users' traces and personal information was already a widespread practice, and it did not stop while users subscribed to their services. The legal framework is such that advertisers do not need to turn to Internet users to access information that they already systematically obtain directly through their websites and applications (data from CRM databases, behavioral data from their website), or indirectly through advertising and marketing players. The founder of Personal, Shane Green, observed in 2013 that the market for personal data "does not exist right now, because consumers are not in on the game." A counterargument might be that it is precisely because consumers are not in on the game that the market *does* exist today (discussed below). In other words, companies like Yes Profile, which promised individuals that they can "own their profile again," are unable to keep this promise: signing up on Yes Profile website and providing one's personal information did nothing to stop the tracking performed by other players (advertising networks, ad exchanges, data brokers, etc.). The profiling performed by Yes Profile only adds to the multiple tracking mechanisms individuals are exposed to, it does not replace them. Consequently, the C2B market for personal data has never really taken off, and the start-ups that have tried to create it are either doing negligible business or have changed their core business, abandoning personal data monetization. Personal fits into the latter category, now selling a collaborative software solution. As for Yes Profile, its web page was inactive in November 2019 and the company seemed to have left the business.

In addition, even if Internet users were to take back control of their data and their traces and try to make money out of them, the self-valuation of personal data by Internet users would come up against another major constraint: on existing markets, individual level information is not worth much, and this effective market price probably does not justify individuals going to the effort of putting them up for sale. Calculations from two sources based on the rates used on the advertising market provide empirical support for this observation. The *Financial Times* was the first, in June 2013 (Steel et al. 2013) to assess the value of typical pieces of personal data (sociodemographic information, assets, leisure pursuits, and consumption patterns) based on list prices from data brokers in the United States. Summing the value of each piece of information, the average total value was running at around 20 cents. A second empirical study by Olejnik et al. (2014) reports purchase prices in online advertising auction systems (known as real-time bidding or RTB). The authors observe that browsing history itself is only worth around $0.0005 for an advertising impression; when applying that price to the average number of sites visited and advertisements displayed per site, the authors estimate the volume of business generated at $0.18 per user per month. Web users' potential earning on this market would be around $0.432 per year.

The failure of the World Economic Forum's forecast of personal data as an "equivalent to 'money'... controlled, managed, exchanged and accounted for just like personal banking services" (Schwab et al. 2011, 10) for individuals can be seen as performative failure of neoliberal ideology. Although personal data can indeed be considered as an asset and a tradable object, the mistake was to consider that its potential value would be unleashed through a C2B marketization process. The idea of personal data as liquidity is also the avatar of the attempt to gauge the potential value of something without being able to dig it into markets. In that sense, the failure of a C2B market for personal data also signals that personal data cannot be considered as a good as-is, but need to be constituted as products to match operational needs and take a specific place in market architectures. In fact, the market for personal data as a product has already existed for decades in the marketing area: a B2B market for purchase intentions, qualified profiles, and lists of individuals, framed around the needs of advertisers and a highly automated industrial supply chain.

The Assetization of Personal Data in the Marketing Industry

The actual market for personal data has its roots before the development of the Internet. It is a B2B affair, with private and public organizations as its main customers, and long-established providers, particularly in the marketing and advertising fields. The activities of marketing and advertising professionals fall into two distinct categories: direct marketing and media advertising. In both cases, the specialists operating on these markets can gain a lot of efficiency in their operations if they have information on how to address their targets. They did not wait for the big data revolution to introduce tools for aggregating, processing, and cross-referencing personal data, turning them into a tradable good between companies. However, the transformations in progress should not be underestimated: we are witnessing large-scale and highly automated industrial engineering of the capture, cross-referencing and use, of individual data. The driving force behind it is the transformation of the central product traded on the advertising market—namely, the audience (see also Turow 2012). In this section, we show that personal data are not traded by data brokers and advertising networks as a commodity, a generic material that would fit a large array of purposes, but rather as refined input tailored to fit the needs of the advertising industry and its products. Personal data are traded in the form of (addressed) contact lists and audience segments, in order to serve as an upgrading and valuing tool for the main products that are traded by advertising and direct marketing firms: ad impressions and direct emails. In that sense, personal data do not constitute a commodified good but rather an asset upon which value can be built, a lever for marginal gains in an already well-equipped market.

Direct Marketing: From List Sellers to Data Brokers

Companies' direct marketing activities target consumers through direct means, as opposed to indirect means (the media). The traditional methods are direct mailing and cold-calling. Over the past twenty years or so, these have been joined by direct emailing. These activities have led to the emergence of an address rental market. This market serves a dual purpose for advertisers, providing them firstly with lists of addresses, preferably up-to-date and with names attached, and secondly with addresses that are qualified—in other words, paired with information that indicates whether or not the concerned individuals are potential purchasers of their products.

The history of address rental, which really took off when companies began to computerize in the 1970s, provides an insight into the place of personal data in today's business practices. Here are the major elements of the story, as told by one of its historic players, French firm ITL: "The early sixties saw many companies specialize in supplying the addresses of people from certain socio-professional categories—doctors, dentists, engineers, students at top universities—collectively referred to, rather eloquently, as high-purchasing-power individuals."[1] In the late seventies, the computerization of small and medium sized enterprises (SMEs) boosted this market through over-the-counter transactions between companies, particularly mail order players, who took the risk of making the names and addresses of their customers available to other companies in exchange for an identical volume of addresses or payment based on a price per thousand.

How did we move from a local and almost artisanal activity in the 1960s to international players building consolidated databases with thousands of variables covering large swaths of the population? Information technology was the key to the change, both for list suppliers and data brokers and for their customers. For the latter, digitization meant not only computerized CRM databases, used to calculate lead scores but also the widespread implementation of loyalty programs, after airplane companies implemented frequent flyers programs in the 1980s (Araujo and Kjellberg 2016) and participated in the spread of this new market device in the retail industry (Pridmore 2008; Coll 2013). Companies thus accumulate knowledge of their customers that includes transactional elements. They can use these elements firstly to get to know their customers better and adapt their pitch and offers accordingly, and secondly to identify prospective customers and customer catchment areas.

Industrial concentration was the main factor leading to the emergence of data brokers—such as Acxiom, Experian, and Epsilon—who built mega databases accumulating as many variables as possible on consumers. The history of the data brokerage companies, as partially unveiled by data brokers themselves (Watson 2013) or academic works (Goss 1995; Hoofnagle 2003; Bouk 2017), shows how today's large players are the result of small companies initially specializing in some particular calculation on individual data: credit scores (Experian, Equifax), political marketing (Acxiom), loyalty programs infrastructure (Alliance Data). As long as these companies needed to have as much customer information as possible, large movements of consolidation led to today's large multipurpose data brokers. Although data brokers are fueled with individual information from various sources, they do not sell personal data themselves but business-oriented informational products based on specific calculations performed on personal data. Their products are lists of qualified leads for direct marketing, marketing segments describing lifestyles and consumption patterns in general terms (e.g., Personicx segmentation by Acxiom), enrichment of existing databases (adding variables to one's customers' CRM bases), and risk calculation (scoring) on individuals or profiles, particularly in the banking and insurance sectors.

Cookies and the Emergence of Data-Driven Advertising

The online advertising market is the other area in which personal data are injected into the value chain and enrich the products exchanged by market players. In the traditional media/advertising economy, the cornerstone of business relations is the audience, which acts as a currency traded between the sector's various players: advertisers purchase access to audiences from the media, and audience measurement institutes (such as Nielsen and Comscore) measure their characteristics and thus their value (Napoli 2003).

A large part of the evolution of the online advertising sector, since its very inception, has been based on taking into account web users' browsing behavior to an ever-increasing degree, so as to qualify and estimate the value of advertising spaces, create products that match consumers' purchase intentions as closely as possible, and make money out of this profile information by selling it to other players in the value chain (Beuscart and Mellet 2013).

Today, the market infrastructure of online advertising—especially its "display" segment, which gathers various forms of advertising which appear on publishers' websites, next to editorial content and account today for about one third of online advertising spending—heavily relies on cookies. Cookies enable a website to store information on the browsers that communicate with it. They are very useful, as they mean a website does not have to rediscover an unknown web user on every page he or she visits. Instead, Internet users can have browsing sessions, store items in shopping baskets, and save their preferences. What we have seen is a redefinition of the audience product based on the tracking capabilities provided by browsing cookies. This evolution was done in three steps.

First, in the early days of online advertising, pricing and sales were based on the model used in the traditional media, with audience measurement used to qualify the profiles of visitors to a site and advertising agencies adding to those qualifications with the content of the pages viewed. The first change cookies made to this model was the possibility for a website and its advertising team to separate visits to sports-themed pages from sales of sports-themed banner ads: by tracking its visitors' browsing over time, a website can now target sports lovers on any of its pages.

As a second step, this first separation, that of visitor qualification from content viewed, was quickly joined by a second on a larger scale: ad networks such as Weborama or Criteo aggregate knowledge of web users' movements on all partner sites—those which allow Weborama to place and then enrich cookies about their visitors. As a result, an advertiser looking for sports lovers can find them on a Weborama-affiliated site whose content has nothing to do with sports: all that is necessary is for the web user to have visited another partner site linked to the sports theme.

The third major change in this market has been the development of programmatic advertising. This is based on the sale of advertising spaces through marketplaces (or ad exchanges), which use auction mechanisms that most often operate virtually in real time, hence the name real-time bidding (RTB). This coordination mechanism automates every aspect of processing: buyers' and sellers' strategies are implemented by algorithms that assess the appropriateness of displaying a banner each time a webpage is opened and seek to optimize the value of such ads. The central element is that these marketplaces are intrinsically designed to put buyers in competition with one another based on the information they have, which once again takes the form of cookies. For all market players, and especially end-buyers (advertisers and media agencies), holding or direct access to cookie bases is a strategic issue in this environment. But not all databases of cookies have the same value: the latter depends on their size (what proportion of the population is covered?), their depth (what is the precision of the qualification of the audience segments?), and their freshness. It is the combination of these three elements that determines the ability of buyers on the one hand to recognize an Internet user when an advertising impression is auctioned and on the other hand to determine the "right" bid.

Thus, in the online advertising market, the browsing data of Internet users recorded in the form of cookies have gradually become more and more important in the daily functioning of the market. For advertisers, cookie databases are accessible to a group of players who have succeeded in taking positions across the web, typically large publishers, ad networks and a few data brokers, usually in the form of buying advertising impressions targeted at a particular audience segment—for example, car purchase intention or female teenagers. The importance of individual data is such that it is reflected in the structure of prices offered by buyers on these marketplaces: at the bottom of the scale, buyers without a cookie who bid for a price close to zero; in the middle, buyers with a third-party cookie recording sociodemographic information or, better, purchase intentions; at the top of the scale, buyers with first-party information—that is, user information that was built in-house. This latter type of information serves as input to retargeting, a display advertising technique used to display advertising to people who have previously visited the advertiser's website; the ad typically displays a picture of a product the visitor was viewing earlier but did not purchase. Retargeting, popularized by start-ups such as Criteo, is the first step in the movement of datassetization that takes first-party data as a base. That is what we are going to see now.

Datassets for Advertisers: The Invention of First-Party Data

Personal data are used in marketing in two historically distinct ways: in the form of contact details and stable and (by and large) verified sociodemographic characteristics in the world of direct marketing; and in the form of cookies aggregating browsing traces and (by and large) predicted sociodemographic characteristics in the advertising world. These two worlds now seem set to converge. But the main instigators of this evolution are not market intermediaries such as data brokers or web services companies; they are the clients of the marketing industry: advertisers.

Data Management Platforms: A Key Infrastructure for the Assetization of Customer Data

The marketing industry has been deeply transformed by digitization; private companies have gone through a similar movement since the 1980s. Private companies have not only introduced computers and databases at any level of their activity, they also are in direct contact with their customers through digital interfaces (billing supports, websites, mobile applications, etc.). Hence, like data brokers or ad networks, they are in a position to collect data. The systematization of tracking, accompanied by the standardization of customer databases, loyalty programs, and online retail, has encouraged marketing teams inside companies to consider personal data as an asset. In the 1990s, a digitized consumer database could be eventually turned to a revenue stream if it was sold to data brokers like Acxiom; in the 2010s, companies have come to consider such information as highly valuable to the point that it should not only be traded but also used as a lever for business action. From companies' point of view, this first-party data, as opposed to third-party data (audience or individual level data available to anyone ready to pay for specific audience segments), is actionable information about their customers they have accumulated and now own in various databases (purchase history, browsing data, response to marketing campaigns, location data, etc.).

Somehow, the designation and consideration of heterogeneous datasets located in various places within organizations as first-party data is characteristic of the implementation of a capitalistic reasoning applied to data. Recently, the main instrument of this evolution has been the introduction and systematizing of data management platforms (DMPs) for advertisers.

In simple terms, a DMP is a database management software that can be used to collect, sort, and analyze data from a variety of sources. The supply of DMPs for marketing purposes emerged in 2011 and has developed rapidly since. The deployment of DMPs is part of a move toward what marketing professionals call "relationship marketing." According to DMP providers, this technology should help companies develop lasting and personalized relationships with their various audiences according to their closeness or distance: customers and prospects (the foundations of customer relationship management) and suspected purchasers or unknown Internet users (the foundations of media advertising). Before DMPs, customer information was already subject to various enrichments with external data (such as segments proposed by data brokers) and calculations (segmentation, scoring) that led to specific actions toward customers (tailored mailing/emailing, special offers), but these actions remained in the field of direct marketing and CRM. DMPs are expected to go beyond this limit, bridging the gap between advertising and direct marketing, by making first-party data actionable in all kinds of direct and indirect actions with customers and noncustomers. Customer data may, for example, be useful in identifying clients and either avoiding their inclusion in a costly media advertising campaign to recruit new clients or, on the contrary, in sending those clients personalized advertising because they are at risk of defecting. The economic consequences are potentially important for companies with a portfolio of several million customers. The DMP is sold as the main cornerstone to drive business actions toward consumers in the wild, and thus as the infrastructure that turns first-party data into assets.

Leveraging First-Party Data: IDs and Algorithms

Yet, leveraging first-party data is not trivial and involves several operations that can be divided in two families: matching and calculation. The DMPs are often described by professionals as "data lakes" merging heterogeneous data sources into one single place. This aggregation process is based on the matching of identifiers among data sources, and in the DMPs, the common identifier is the cookie. Customer databases typically use the name, email address, or phone number as an identifier while the cookie serves as an identifier to trace the browsing of Internet users. This means it is necessary to "cookify" customer data in order to onboard it in the DMP. To do so, companies can encourage customers to log in online, so that it is possible to match an email address with a web cookie. For example, the French retailer Carrefour is forcing its customers to activate their loyalty card on their website, thus setting a cookie on their computer containing their loyalty card information, which will be reused for any further visit. Another method is to use a third-party actor specializing in onboarding, who is responsible for finding bridges between lists of addresses or phone numbers, and cookies. These specialists are developing partnerships with websites where users complete forms, such as insurance products comparators. Companies such as LiveRamp (acquired by the data broker Acxiom in 2014) are specialized in this niche. The onboarding operations of customer data are only one type of basic matching operations among others performed within or besides the DMP. Once transformed into cookies, the customer data must then be matched to the databases of the advertising partners from which the advertiser chooses to launch these campaigns. The multiplication of these operations of onboarding, matching, and synchronization also results in losses and in the end, the cookie lakes feeding the DMP often cover a limited part of the advertisers' effective clientele.

The other way to leverage first-party data is to perform computational operations on the data. These operations rely on algorithmic apparatuses. The objective here is to identify specific audience segments within the DMP and in connection with external web user information from third-party services, based on statistical proximity. A very common operation for advertisers today is to locate a segment of highly profitable customers in their client base, run algorithms to analyze their characteristics, and then launch advertising campaigns to nonclient audience segments with the same characteristics. These so-called lookalike segments are created by ad networks or big players such as Facebook and sold in the form of advertising campaigns.

Based on these two techniques, matching and calculation, first-party data are made expandable and can combine with all kinds of third-party data. The DMP is the base of these processes within organizations. It is literally, and concretely, both a data infrastructure (it is a database, or a second-level database: a database made of databases) and a valuation device (it is a knowledge device but also an action tool and a measurement tool that very precisely assesses the efficiency of different marketing actions). It is not purely anecdotal that the metaphor of the bank, used by the World Economic Forum to describe the mechanisms of personal data storage and marketing by individuals themselves, is reemerging, this time about DMP. In the words of eMarketer, a reference market research company in the field of digital marketing: "Data is the currency of today's digital advertising ecosystem, and everything from media planning to analytics—and every campaign execution in between—runs on it. If data is the currency of digital marketing, then the DMP is the bank" (Fisher 2013). The metaphor is not particularly detailed in the report, but the use of the image of the bank refers to that of the institution of capitalization par excellence.

Conclusion: Assets as Market Infrastructure

Today, personal data are widely understood as the counterpart of free online services in the mainstream media, with the usual claim "if it's free, you are the product," and in the academic field as well (Carrascal et al. 2013). But with a closer look at how personal data are forged with tracking and collection technologies, how they circulate within the advertising industry, and how they become part of products within this industry, it appears that personal data are not marketized as the product in this market, although they came to constitute a key element in its architecture. Our inquiry into the integration of personal data within the industry of marketing allows us to highlight three aspects of assetization that are more general in scope than the case described in this chapter.

First, assetization, as a process, is a combination of capture and repurposing. Interestingly, public debates on already assetized things typically focus on how firms capture things through a combination of legal and socio-technical tools to ensure rentiership from these assets (Birch 2019). For instance, Ebeling (2016) shows that private healthcare companies have elaborated sophisticated arguments to claim that patient data need to be considered as their own property. Still, this protective move comes second in the process of assetization. Before that, data-things must be considered as valuable, and this implies that they are necessarily repurposed—transposed from a specific scene to a different scene where they can be valuated and eventually traded. On this basis, as revenue streams become real, their technical and legal frame is consolidated, and supplementary revenue streams are ensured through specific infrastructures that can almost cannibalize the original "things." Cookies are a good example of the infrastructural dimension of assetization: a simple mechanism initially aimed at filling online baskets has little by little been invested and overloaded by marketing professionals, who have built upon it complex and intricate market data-based infrastructures for data-related goods (ad exchanges, real-time bidding, etc.). Eventually, the cookie has been partly cannibalized by the ad tech perspective, which turned it into its key infrastructure for the capitalization of personal data (Mellet and Beauvisage 2019).

Second, assetization can be considered as an entrepreneurial work, especially in the repurposing stage of the assetization process. Although unsuccessful, the attempts to create a C2B market for personal data are exemplary of the dynamics of start-up entrepreneurship in the digital economy. Another example is the way the first ad networks convinced numerous websites to put tracking pixels on their pages in exchange for audience statistics, or how today firms invest in data management platforms software as a potential revenue stream for the future. In all cases, assetization is a dynamic process of exploration driven by entrepreneurs as market intermediaries and makers, with uncertain and unstabilized future revenue streams.

Finally, once stabilized, the resulting status of assets is plural and versatile. Within the online advertising market, personal data are considered as informational goods when data brokers complement consumer bases with their own information sources; they are market-based matching devices when advertisers chase their own customers on Facebook to address them with tailored messages; they are essential qualities for the valuation of ad inventories for online publishers and ad networks. In all matters, they have become essential elements in the digital advertising industry, and the common claim that digital advertising is selling user data is, although not correct in most cases, a good synecdoche for what digital ad inventories have become. However, the assetization process has led firms to consider personal data as a valuable asset for their own purpose, coining them as "first-party data." These untraded assets have more to do with global efficiency and work instruments for marketing operations. This plurality of economic statuses demonstrates the versatile and plural nature of assets.

Note

1. ITL. Petit historique du marché de la location d'adresses. Corporate document, http://www.itl.fr/en/fichier/59-petit-historique-marche-location-adresses.

References

Araujo, L., and Kjellberg, H. 2016. Enacting Novel Agencements: The Case of Frequent Flyer Schemes in the US Airline Industry (1981–1991). *Consumption Markets & Culture* 19 (1): 92–110.

Beresford, A. R., Kübler, D., and Preibusch, S. 2012. Unwillingness to Pay for Privacy: A Field Experiment. *Economics Letters* 117 (1): 25–27.

Beuscart, J.-S., and Mellet, K. 2013. Competing Quality Conventions in the French Online Display Advertising Market. *Journal of Cultural Economy* 6 (4): 402–418.

Birch, K. 2019. Technoscience Rent: Toward a Theory of Rentiership for Technoscientific Capitalism. *Science, Technology, and Human Values*, doi:10.1177/0162243919829567.

Bouk, D. 2017. The History and Political Economy of Personal Data over the Last Two Centuries in Three Acts. *Osiris* 32 (1): 85–106.

Carrascal, J. P., Riederer, C., Erramilli, V., Cherubini, M., and de Oliveira, R. 2013. Your Browsing Behavior for a Big Mac: Economics of Personal Information Online. In *Proceedings of the 22nd International Conference on World Wide Web*, 189–200. ACM.

Christl, W., and Spiekermann, S. 2016. *Networks of Control. A Report on Corporate Surveillance, Digital Tracking, Big Data & Privacy*. Vienna: Facultas.

Coll, S. 2013. Consumption as Biopower: Governing Bodies with Loyalty Cards. *Journal of Consumer Culture* 13 (3): 201–220.

Danezis, G., Lewis, S. and Anderson, R. J. 2005. How Much Is Location Privacy Worth? In *WEIS* 5.

Ebeling, M. 2016. *Healthcare and Big Data*. New York: Palgrave Macmillan.

Eubanks, V. 2018. *Automating Inequality: How High-Tech Tools Profile, Police, and Punish the Poor*. New York: St. Martin's Press.

Federal Trade Commission. 2014. *Data Brokers: A Call for Transparency and Accountability*. Washington, DC: Federal Trade Commission.

Fisher, L. 2013. Data Management Platforms: Using Big Data to Power Marketing Performance. EMarketer Report.

Fourcade, M., and Healy, K. 2017. Seeing like a Market. *Socio-Economic Review* 15 (1): 9–29.

Goss, J. 1995. "We Know Who You Are and We Know Where You Live": The Instrumental Rationality of Geodemographic Systems. *Economic Geography* 71 (2): 171–198.

Hoofnagle, C. J. 2003. Big Brother's Little Helpers: How ChoicePoint and Other Commercial Data Brokers Collect and Package Your Data for Law Enforcement. *North Carolina Journal of International Law and Commercial Regulation* 29: 595–637.

Huberman, B. A., Adar, E., and Fine, L. R. 2005. Valuating Privacy. *IEEE Security Privacy* 3 (5): 22–25.

Kitchin, R. 2014. *The Data Revolution: Big Data, Open Data, Data Infrastructures and Their Consequences*. London: Sage.

McFall, L. 2014. *Devising Consumption: Cultural Economies of Insurance, Credit and Spending*. London: Routledge.

Mellet, K., and Beauvisage, T. 2019. Cookie Monsters: Anatomy of a Digital Market Infrastructure. *Consumption Markets and Culture*, 1–20.

Milyaeva, S., and Neyland, D. 2016. Market Innovation as Framing, Productive Friction and Bricolage: An Exploration of the Personal Data Market. *Journal of Cultural Economy* 9 (3): 229–244.

Muniesa, F., Doganova, L., Ortiz, H., Pina-Stranger, A., Paterson, F., Bourgoin, A., Ehrenstein, V., Juven, P.-A., Pontille, D., Saraç-Lesavre, B., and Yon, G. 2017. *Capitalization: A Cultural Guide*. Paris: Presses des Mines.

Napoli, P. 2003. *Audience Economics: Media Institutions and the Audience Marketplace*. New York: Columbia University Press.

Nissenbaum, H. 2017. Deregulating Collection: Must Privacy Give Way to Use Regulation? SSRN, https://ssrn.com/abstract=3092282.

Olejnik, L., Tran, M.-D., and Castelluccia, C. 2014. Selling Off Privacy at Auction. In *Proceedings 2014 Network and Distributed System Security Symposium*. San Diego: Internet Society.

Posner, R. 1981. The Economics of Privacy. *The American Economic Review* 71 (2): 405–409.

Pridmore, J. 2008. Loyal Subjects? Consumer Surveillance in the Personal Information Economy. PhD Thesis. Kingston, ON: Queen's University.

Schneier, B. 2015. *Data and Goliath: The Hidden Battles to Collect Your Data and Control Your World*. New York: Norton.

Schwab, K., Marcus, A., Oyola, J., Hoffman, W., and Luzi, M. 2011. *Personal Data: The Emergence of a New Asset Class*. Switzerland: World Economic Forum.

Staiano, J., Oliver, N., Lepri, B., de Oliveira, R., Caraviello, M., and Sebe, N. 2014. Money Walks: A Human-Centric Study on the Economics of Personal Mobile Data.

In *Proceedings of the 2014 ACM International Joint Conference on Pervasive and Ubiquitous Computing—UbiComp '14 Adjunct*, 583–594. Seattle: ACM.

Steel, E., Locke, C., Cadman, E., and Freese, B. 2013. How Much Is Your Personal Data Worth? *Financial Times*, June 12, 2013.

Turow, J. 2012. *The Daily You: How the New Advertising Industry Is Defining Your Identity and Your Worth*. New Haven, CT: Yale University Press.

Watson, N. 2013. *A Brief History of Experian. Our Story*. Dublin: Experian.

4 A Crisis for Cures? Tracing Assetization and Value in Biomedical Innovation

Victor Roy

Introduction

New advances in biology—from the rise of genomics in the late twentieth century and more recent breakthroughs in gene editing technologies—have conjured the cultural imagination of the "cure." When Mark Zuckerberg and Priscilla Chan launched the Chan-Zuckerberg Initiative in 2016 with their bounty from Facebook, they put forward an audacious goal: "to cure all diseases within the century" (Sample 2016). In the same year, representatives from both political parties advanced and eventually passed the 21st Century Cures Act in the US Congress, with the explicit aim of accelerating more cures to patients (Hudson and Collins 2017). Part of this legislation supported the "Cancer Moonshot" led by Joseph Biden, which carries the aim of delivering long-sought after cures for the elusive scourge (Lowy and Collins 2016). If the prior century was a golden age for physics, our current century has been branded by prominent scientists as one defined by the young and maturing field of biology and its promise for human health (Venter and Cohen 2014).

This chapter locates these hopes, however, within a context of a paradox: just as these imaginations are gaining greater traction, buttressed in part by gains in the science of human disease and treatment, a future of curative therapies appears further out of reach. To unpack this contradiction, I explore recent transformations in political-economic dynamics and valuation practices in biomedicine that can belie the very existence of curative therapies. These transformations—which place what Birch and Muniesa (this volume) describe as the "asset form" at the center of scientific and technological innovation for health—combine to constitute what I call the *assetization of biomedicine*. This analysis departs from more commonly used lenses within sociological and anthropological traditions that have focused on the processes and impacts of *commodification* on healthcare—especially the consequences of exchange relationships in capitalist markets on the objectification of illness (Sharp 2000). By contrast, to comprehend the ownership relationships and future-oriented subjectivities to human health that I argue are central to the unfolding of contemporary biomedicine, this chapter charts an alternative analytic path.

To understand the nature of the paradox, then, I draw on scholarship from science and technology studies and political economies of innovation to define two central turns in the late twentieth century that have underpinned the assetization of biomedicine (Andersson et al. 2010; Birch 2017; Dumit 2012a; Lazonick and Tulum 2011; Mazzucato 2016; Pisano 2006). The first turn has been the emergence of a "relay race" model of drug development in which ownership over intangible knowledge assets is passed along multiple actors (i.e., venture capitalists, small biotechnology companies, and the shareholders of large, publicly traded companies) from early stages to regulatory approval (Lazonick and Tulum 2011; Pisano 2006). Rather than being used to recoup research investments within integrated pharmaceutical companies, patents in this model take on a new function as monetized assets. Monetized assets in these financial markets are valued not for their current profitability but for the downstream earnings potential they bring. This first turn has been codependent with a second: an evolution in the epistemic practices and narratives of valuation within clinical medicine and health economics in which health is abstracted away from the felt illness experienced by patients toward a quantified valuation of statistical health improvements and averted future disease at the level of populations. In the process, illness is redefined as risk, and health as future-oriented risk reduction (Dumit 2012a). Taken together, the assetization of biomedicine denotes both the political-economic transformations involved in making knowledge into assets for financial valuation and business strategies, and also the associated epistemic, narrative transformations that turns health into a future-oriented "fact."

Studying assetization thus reveals the ways in which the structures and practices of valuation in finance become entangled with the valuation of health by patients, physicians, and policy makers. To examine this entanglement and its consequences, curative therapies are a particularly interesting class of "things" to consider. By ending disease processes rather than mitigating their progression, curative therapies are culturally and biomedically valorized precisely because in producing positive health outcomes, they obviate their own necessity. Yet due to the transformations involved with assetization, curative therapies are caught within two vexing crosswinds.

In one crosswind, cures carry the possibility of eliminating the very potential for continual revenue growth on which their value as assets in financial markets rely. Cures thus present a structural business challenge for biotechnology and pharmaceutical companies (Dumit 2012b). In another crosswind, the rare curative therapies which *are* developed are valued as assets that create seemingly infinite possibilities for health improvement, which is then used as justification for increasing financial remuneration to their manufacturers (Reinhardt 2015). This remuneration, represented in the high prices of new therapies, ultimately challenges governments and health systems with finite budgets and limits patient access. Curative therapies are thus within tantalizing reach for many diseases due to advances in basic and clinical science, but what can we make of their future against the terrain of the assetization of biomedicine? After describing the transformations that define the assetization of biomedicine, I use the case of curative therapies to explore the central logics and limits of assetization and conclude by considering the competing dynamics that could shape whether this cultural imagination of curative therapies can one day mirror reality for patients.

The Context: Commodification's Blind Spots as an Analytic Frame for Biomedicine

The need to bring the "asset" form into an analysis of biomedicine, however, may not at first blush be obvious. Indeed, sociologists and anthropologists have produced a rich tradition of analyzing biomedicine, and the relationships between the "things" of biomedicine—such as drugs, genes, bodies—and their social contexts (i.e., people, markets, institutions). A central thread in this tradition has been unpacking the implications for the configuration of these "things" as commodities which, through their entry into exchange relationships in capitalist markets, become emblematic of multiple transformative social processes captured under the banner of "commodification" (Csordas 1994; Pellegrino 1999; Scheper-Hughes and Lock 1987; Taussig 2010; also see Braun, this volume).

Of concern in this literature is tracing how the exchange relationship in capitalist markets renders human categories—such as the moral and phenomenological experience of health and illness—into an object of economic desire (Pellegrino 1999; Sharp 2000). The objectification in this view reduces human life to a transaction, where the value of health (represented in a drug, intervention, or surgery, for example, necessary to sustaining life) is exchanged for some counterpart value—most often indirectly, through a system of money. This process has been interrogated, often from a vantage of critique, across a wide array of settings and cases—such as with the privatization of health services due to neoliberal structural adjustment policies in developing countries, or the proliferation in technological intervention (i.e., medicalization and pharmaceuticalization) on human bodies across industrialized countries in the postwar era (Keshavjee 2014; Abraham 2010).

A related but distinct concern of commodification studies that focus on biomedicine has been the intertwining of more recent advances—such as gene-editing and reproductive technologies—with new iterations of financial capital and markets. The angle here has been to examine the impact of these technologies in creating new biological sites (i.e., genes, organs, embryos) for capitalist accumulation and thereby further fragmenting the body (Helmreich 2008; Sharp 2000). Studied with the popularized conceptual apparatus of "biocapital," scholars have been particularly adept at exploring the ways in which technological advances and financial capitalism enables biological materiality to enter into commodity relationships, as well as the critical function of promissory rhetoric in mobilizing speculative capital to convert this materiality into capitalist value and accumulation (Helmreich 2008; Rose 2007; Sunder Rajan 2006).

These literatures, while offering valuable insights into the intersection of capitalism and biomedicine, leave open at least three directions for investigation that are particularly relevant in studying the possibilities for curative therapies. First, as Birch (2017) has detailed, discourses of the future and the conversion of biological materiality via the commodity form do not sufficiently capture the genesis and management of value in these modes of accumulation. Extending our grasp of value in contemporary biomedicine requires interrogating the political-economic and valuation practices that *make* this conversion and discourse possible. By this view, cures are not inevitable, based on some biologically latent potential or promissory rhetoric, but are mediated by social dynamics that shape their prospects.

Second, while a concern with commodities has focused on exchange relationships, a consideration of the asset form allows a focus on the proliferation of *ownership relationships* in biomedicine—such as with the control of intellectual property over scientific and therapeutic knowledge (Birch 2017; Gagnon 2016). This control is, as highlighted in the introduction to this volume, also linked to market logics in financial markets that are different than those of commodities—where value is tied less to profitability but more to the potential stream of earnings that ownership of an asset may bring. This future-orientation has implications for cures which may be profitable but by definition can carry uncertainty for ongoing accumulation.

Third, this configuration of ownership, centered on future-oriented valuation, can lead to financial modes of value which colonize social spaces beyond markets—from institutions such as health delivery systems to the doctor-patient relationship, to individual perceptions of disease (Chiapello 2015; Van der Zwan 2014). Though capitalist relations can alienate and objectify through the commodity form, an analysis of the asset form may reveal the ways in which common understandings of health may also be reoriented toward an "investment" centered aspiration—where health too is an asset, and where the quantification of this value (signaled through prices) aims to legitimate and valorize. This emphasis on the future can translate to new kinds of *subjectivities* toward health among patients, scientists, public officials and business executives. The realization and affordability of curative therapies, in turn, may depend in part on the epistemic practices and narratives of value that reify these subjectivities. These relative blind spots in commodification studies regarding biomedicine present an opportunity for inquiry. My task in the remainder of this chapter is to trace the asset form and its implications for biomedicine: to do this, I look at the case of biomedical innovation—the process by which new therapies are developed—and the two transformations that have been at the heart of its assetization.

Patents for Research to Monetized Patents: Turning Drug Development into a Relay-Race of Assets

The first turn has been in the political-legal and economic dynamics underpinning biomedical innovation, where value is derived less from the revenue and sales of commodities—such as approved medicines—than from capitalization and control over intangible assets in financial markets. In this model, therapeutic development has become akin to a relay race, in which intangible assets (i.e., knowledge property) are passed along multiple actors—from the state to venture-backed companies to large biopharmaceutical companies (Birch 2017; Pisano 2006). While this model has attracted an explosion of capital and served as the basis for significant sums of capital accumulation, it has also exposed its actors (especially the "anchor" in the relay race—large, publicly traded pharmaceutical companies) to a structure of vulnerability and crisis that has implications for the fate of curative therapies (Sunder Rajan 2012).

For much of the postwar era of biomedical innovation—which witnessed the ascendance of large industrial pharmaceutical companies—research was carried out within the confines of large firms as well as publicly supported labs (Slaughter and Rhoades 1996). To attract investments for research and development, intellectual property protections granted to these firms by the state were viewed as the critical incentive (Scherer 2004). This patent centered organization of biomedical research has been understood to facilitate scientific development, in which inventors receive monopoly ownership for a specific period of time (twenty years from the time of invention) only after which the public can gain full access (i.e., generic licensing) to the knowledge produced by the patent. As the STS scholar Biagioli (2006) has chronicled, patents have been viewed as governing a legal exchange between consumers and investors of patent protected products, with these transactions conceptualized in the law as a "bargain" or "fair exchange"—investors' right to recuperate costs of research and development in exchange for customers' access to the inventor's product.

The importance of this argument to the pharmaceutical industry—in which monopoly pricing is justified on the basis of research and development costs—is illustrated by a series of studies by the industry-funded Tufts Center for Drug Development. Since the 1980s, economists there have published estimates of the escalating costs of research and development (upwards of $2.7 billion per drug as of 2014) as an illustration of the need to maintain intellectual property protections and the significant pricing power these protections grant (DiMasi et al. 1991; DiMasi et al. 2016). In this conception, the prices that companies later charge for approved, patented medicines is, in turn, ostensibly tied to the investments made within the walls of corporate laboratories. Though the DiMasi figures have been disputed (with alternative studies showing much lower costs for research and development), this justification for patents and prices has continued to dominate folk understandings of biomedical innovation (Light and Warburton 2011; Nik-Khah 2014). Yet even as this ongoing debate over patents and the costs of research and development has persisted, a series of changes around 1980 led to a transformation in the actual function of patents in biomedical innovation.

With a flagging economy and facing new global competition from Europe and Japan, US policy-makers viewed government funded research in the 1970s as a previously untapped source of growth (Rai and Eisenberg 2003; Slaughter and Rhoades 1996; Vallas et al. 2011; Berman 2012). To pursue this direction, the US Congress passed the Bayh-Dole Act in 1980s, which allowed for the private patenting of government-funded research for the purposes of commercialization (Rai and Eisenberg 2003). In the decade that followed, universities across the country developed technology transfer offices, designed to support researchers to convert their publicly funded projects into commercial ventures (Kenney and Patton 2009; Mowery and Sampat 2004). With the opportunity to gain equity in start-up enterprises (and with corporate bureaucracies more risk-averse in developing and adopting new technologies), professors could view their research as holders of business potential (Block and Keller 2009; Berman 2012). Some universities could, in turn, make handsome gains through royalty agreements with these new enterprises (Mowery and Sampat 2004).

The financing of these new firms would come from a nascent source emerging during the 1970s and 1980s: venture capital. Borne in part from changes in pension regulation in the 1970s that enabled pension funds to direct more of their capital to riskier ventures as well as new technological opportunities (e.g., computing) that brought down the start-up costs associated with new companies, venture capitalists pursued investments in start-up enterprises as a vehicle for financial returns as well as industrial transformation (Gompers 1994; Gompers and Lerner 2004). The early model from that period remains today: venture capitalists typically provide rounds of capital in the early stages of a company's development in exchange for an ownership stake in the company (Robbins-Roth 2001). Viewing themselves as "active investors," venture capitalists join the boards of the company, attempting to use their networks and financing to convert uncertain enterprises into high value investments (Hopkins et al. 2013). In this model of investment, the exit is of pivotal importance, as venture capitalists typically stick around for three to five years, with initial public offerings on equity markets (e.g., NASDAQ) or acquisitions by larger companies as the pathways for generating returns (Robbins-Roth 2001). Advances in molecular biology in the 1970s along with the Bayh-Dole Act attracted venture capitalists to the emerging sector of biotechnology during the 1980s and 1990s (Pisano 2006). The early success of companies like Genentech stoked the hopes that new enterprises could convert these new advances in science and push the frontiers of a so-called biotech revolution (Nightingale and Martin 2004).

These new venture-backed biotech businesses exemplified a shift in business strategy. As documented by management scholar Gary Pisano (2006), these businesses and their venture backers aimed to monetize intellectual property—a stark departure from integrated, incumbent firms that sought to bring drugs from laboratory testing all the way to patients. These biotechnology companies rarely bring drugs to market and are not valued for revenues on sales; rather, their intangible assets are valued based on the earnings stream they might one day bring. This has led to the phenomenon of what Lazonick and Tulum (2011) have called "product-less IPOs"—with most biotechnology companies having no approved drugs at the time of their public offering. The existence of financial markets, however, provides investors and traders opportunities to gain a return by entering and exiting their stake in these enterprises, either through a sale, public offering, or fluctuations in share price. In this conception, patents are not bargains between investors and the public for recouping the costs of research and development, but rather tied to the expectations of future value materialized in financial markets.[1]

While biotechnology companies entered onto the scene during the 1980s and 1990s, large, incumbent pharmaceutical companies were undergoing a transformation of their own. This period witnessed the rise of shareholder value-maximization (SVM) as a corporate governance strategy with profound implications across the US economy—including the pharmaceutical sector. In this latest iteration in a long-running debate over who should control the firm, shareholders, not managers, were viewed as efficient allocators of capital in the economy. Scholars from law, economics, and finance advanced the notion that shareholders could use share prices in financial markets to direct capital toward higher growth companies and sectors, whereas managers might instead allocate capital toward increasing the size of their businesses or risky projects that might be wasteful (Fama and Jensen 1983; Jensen and Meckling 1976). Additionally, shareholders were viewed as the only economic actors in corporations who make productive contributions without a guaranteed return—as opposed to creditors, workers, suppliers, and distributors, who are compensated through a market-determined price for goods and services (Fama and Jensen 1983). This claim has been used to justify shareholders as the recipients (through dividends and share buybacks) of any "residual" profit left over after a company has paid all their other stakeholders (Lazonick 2015). Taken together, SVM has meant directing accumulation to shareholders—both because of their role in the broader economy (as "efficient allocators") as well as their roles within corporations (as "residual claimants"). To discipline executives to follow this shareholder mandate, corporate boards and shareholders have made executives into major shareholders by offering generous stock-based rewards that tie their compensation to share price performance (Lazonick 2015).

In this context of SVM, large pharmaceutical companies are evaluated not on their current profitability, but on their potential to deliver future earnings growth for shareholders. For the incumbent companies with existing streams of revenue, shareholders expect 8 percent to 10 percent returns on an annual basis. This figure mirrors the cost of capital, which roughly represents the returns on capital that investors can expect to receive in financial markets from ownership in other assets (such as mutual funds and bonds) (Damodaran 2017; Nitzan and Bichler 2009). This shareholder-driven growth, however, exposes these larger, publicly traded companies to what Sunder Rajan (2012) has described as a two-sided structural crisis.

On one side of this structural crisis, companies face "patent cliffs"—the expiration of intellectual property protections on their existing assets that bring streams of revenue. With expirations, incumbent firms lose these streams to generic manufacturers, which can charge much lower prices—a margin on top of manufacturing costs—for their unbranded products. Take the example of Pfizer, which lost protections for its cholesterol-lowering drug Lipitor in 2011—and thus an earnings stream of more than $10 billion, accounting for over 20 percent of the company's total revenue (Harrison 2011). Between 2010 and 2014, the industry experienced an erosion of an estimated $78 billion in sales due to patent cliffs (Harrison 2011). Replacing these streams of revenue—*and* generating growth—can require bringing major new therapies to market almost every year. But this requirement collides with the realities of biomedical innovation, where product development often takes over a decade and is riven with failures, due to the risky nature of human clinical trials which other sectors (such as information technology) generally do not face (Pisano 2006).

This time horizon for investment takes us to the other side of the structural crisis: dry pipelines of potential drugs within large, incumbent pharmaceutical companies. Faced with meeting near-term growth expectations, executives of large companies are more risk-averse to the longer-term, patient investments needed to stock pipelines through early-stage science and preclinical testing. Upon the expiration of Lipitor, for example, Pfizer had no drugs in development that could replace the lost revenue (Harrison 2011). This lack of investment does not signal a lack of resources but rather the influence of SVM as a corporate strategy. After shutting down its anti-infective research unit less than two years before, for example, Pfizer spent the first three quarters of 2015 directing $11.4 billion in share buybacks and dividends to its shareholders (Roy and King 2016). Dry pipelines are one outcome of this aversion to relative long-term investment, itself a product of shareholder control in financial markets.

To meet the shareholder expectations of growth while facing patent cliffs and dry pipelines, large companies have become almost structurally positioned in the innovation process as acquisition and late-stage clinical trial specialists, betting on therapeutic assets that may be potent near-term vehicles for earnings growth. Pharmaceutical and life sciences companies spent $228 billion on mergers and acquisitions in just the year 2015, illustrating the reliance on such transactions to generate growth. Gilead's former CEO John Martin, sharing a financial market friendly view held among many pharmaceutical executives, reassured investment analysts on a 2015 earnings call by saying, "we typically like things where we can have an impact on phase III," indicating the later-stages in drug development where companies tend to prefer acquisitions to generate near-term growth (Roy and King 2016). In making these acquisitions, larger companies like Gilead use "capitalization" as a calculative device through which to forecast this potential accumulation. As part of this quantification, companies anticipate the prices they can charge health systems—a topic to which we turn in the next section—and use debt and stockpiles of capital from previous sales to make what are often sizable bets on new streams of earnings.

This organization of innovation carries dramatic implications of the kinds of drugs that are developed. Larger companies—possessing the comparative advantages of global regulatory, manufacturing, and distribution expertise that smaller companies lack—become gatekeepers for the kinds of drugs that make it to patients. In this calculus of growth, the smaller companies (either venture-backed or publicly traded after an IPO) described earlier have become developers and suppliers of assets that larger, incumbent companies will value (Andersson et al. 2010). Acquisitions and late-stage clinical trials by these larger, incumbent companies in turn have become almost entirely oriented around meeting the continual requirement of accumulation and growth, which has quite apparent consequences for curative therapies.

This asset-centric, relay-race model of drug development represents the political-economic transformations that underpin the assetization of biomedicine. Intangible assets serve as the basis for valuation along this relay race and are configured not as patents for recouping research investment, but as monetized property that become vehicles for accumulation for multiple economic actors (from venture capitalists to equity traders) in financial markets. The future-oriented ontology of assets here shapes the trajectory of biomedical innovation, where profitability and revenue from sales of existing products is less the focus: at stake is betting and trading on quantified expectations of the future. This ontology of assets not only privileges certain kinds of assets—those that can generate continual accumulation—but it is also intertwined with changes in our understanding of health and illness.

From Felt Illness to Assetized Health: The Epistemic Practices of a Future-Oriented Health

Generating continual accumulation in financial markets has required a second critical turn defining the assetization of biomedicine: a changing locus of value in health, in which health itself is configured and valued as an asset. Health here is not viewed as healthiness, which requires no therapeutic intervention but rather as a state of preventing potential disease through ongoing and growing therapeutic consumption. Arriving at this understanding of health and the role of treatments has required the development of two sets of epistemic practices—one within clinical medicine and epidemiology and another in health economics—which has in turn buttressed a narrative transformation of health. In this newer narrative, health through treatments is valued by individuals and patients for reducing risk and valorized by drug developers for their potential in creating population health improvements. Before unpacking the implications of this version of health for the materialization of curative therapies, I elaborate these shifts in epistemic practices and narratives.

The first set of epistemic practices has been described by Joseph Dumit (2012a), where the rise of prospective clinical studies and clinical trials in the postwar era enabled a redefinition of health from a binary of healthy/sick toward a continuum of risk. Much of the early history of modern biomedicine in the nineteenth and twentieth centuries had centered on a "felt illness" model of disease, which had in turn framed the need for new treatments: put simply, treatments allow people who feel sick to get better. In this prevailing scenario, physicians would attempt to provide the antidote for a patient seeking care for an episode of illness.

Yet contemporary biomedicine has been demarcated by a major change: alongside "felt illness" models of disease a statistical, population-level model of disease has emerged in which health is realized by the extent to which the risk for potential downstream disease is reduced. In Dumit's tracing of this empirical phenomenon, a central route to achieving this risk reduction is ongoing diagnosis and treatment, which, from the vantage of business, has become a primary vehicle for accumulation. To pharmaceutical companies, patients become valuable only when they can consume more treatments. Dumit (2012a, 17) calls this phenomenon "surplus health," which he defines as "the *capacity* to add medications to our life through lowering the level of risk required to be 'at risk.'" This capacity has been shaped by the rise of two developments in clinical medicine and epidemiology: prospective clinical studies and clinical trials.

By tracking individuals over long periods of time and examining the links between biological markers (such as cholesterol or blood pressure) to downstream events (like heart attacks and mortality), prospective clinical studies rendered visible a phenomenon previously shrouded: the risk of disease progression. The Framingham Heart Study, begun in 1950, is the most prominent early example of such a study, tracking over 5,000 members of a small town over multiple generations to uncover the risk factors of heart disease (Dumit 2012a). Through measuring statistical links between risk factors and disease, such prospective clinical studies enabled what doctor-epidemiologist Geoffrey Rose (2008, 42) described as "a type of disease not hitherto recognized in medicine in which the *defect is quantitative not qualitative* (emphasis added)." Unbeknownst to the patient, even small imbalances (such as high blood pressure or low thyroid levels) could indicate the early onset of disease. The downstream impact of these small imbalances, in turn, is detectable only by measuring large populations over long time horizons. From diabetes to heart disease, from psychiatric illness to Alzheimer's disease, such clinical studies have served to reframe our conception of disease from binaries (healthy/sick) to a continuum of risk.[2]

To intervene on this continuum of risk required another development in the postwar era: randomized clinical trials. By comparing treatment arms versus control arms (that received the prior standard of care or placebo) in samples deemed to be representative of populations, randomized clinical trials aggregated evidence about the effects of treating individuals on the health of populations. Such evidence has produced what Dumit calls an array of "public health facts" (e.g., taking X medication for five years reduces the chance of heart attack by 20 percent), where once a patient crosses a risk threshold (as indicated by clinical studies and trials), treatment becomes the primary way of reducing the probability of future disease. Detecting these population-level effects required repeat treatment over long periods of time in large numbers (often hundreds or thousands) of individuals. Through this process, large-scale trials redefined disease from episodic states of abnormality in need of acute treatment to chronic events demanding long-term management and prophylactic interventions to be deployed even before the onset of disease (Sunder Rajan 2012).

For drug developers and pharmaceutical companies, this redefinition of health with clinical studies and trials became fundamental to the business strategies of continual capital accumulation required by financial markets. In the view of drug developers, new market potential was unlocked in two related ways. First, every person—even those not experiencing any felt illness—could now become subjects for intervention by crossing certain quantitative thresholds (treatment eligible). Companies could find, as one executive put it, "more hidden patients among the apparently healthy" (Dumit 2012a, 114). This meant larger market size. Second, viewing health on a continuum also grew the duration of treatment that patients might require, with disease mitigation requiring years or even lifelong treatment.

The development of statins and anti-diabetic drugs along with the aggressive pursuit of anti-Alzheimer drugs are examples of such regimens. As Dumit (2012a, 115) puts it, "Diseases previously regarded as incurable downward progressions came to be seen as long-term chronic conditions requiring prediction, surveillance, and chronic treatment." Mitigating disease through chronic treatment, then, could become a powerful accumulation strategy for companies in pursuit of continual growth.

But this mode of accumulation has been dependent not only on manufacturers seeing this "surplus health" as a locus of value, their customers—health systems around the world—have also needed to behold this value and be willing to pay for it. Translating risk reductions through treatments into a durable mode of accumulation has thus required another pivot: the economic quantification of this future-oriented value, in which the price and costs of therapeutic interventions are deemed commensurate with the value of health improvements for health systems and the populations for which they are accountable. Alongside developments in clinical medicine and epidemiology, the emergence of health economics thus contains the second set of epistemic practices of importance to the assetization of biomedicine.

With the prices of new medicines typically multiples above the median wages of individuals, assessing their value has fallen to the ultimate customers across the world: public health systems and private insurers, depending on the given country (Reinhardt 2015). With limited budgets, these buyers make pivotal determinations over how to generate the most health improvement for their populations with the money they have. To align their business strategies to the preferences of their buyers, companies which typically had defended their prices on the costs of innovation have turned to a different strategy called "value-based pricing" (Claxton et al. 2008; Gregson et al. 2005; Maldonado Castañeda 2016).

In this strategy, manufacturers set prices based on the "value" that they anticipate health systems will attribute to a given therapy. Manufacturers make their estimates based on their knowledge of the health economics methodologies that health systems use in assessing this value. Through "cost effectiveness" research, for example, health systems weigh the future benefits versus costs of a given treatment strategy (Claxton et al. 2008; Weinstein et al. 2009). New treatments, in this research, are tested for whether they create more health in the future—measured through a unit of health known as "quality adjusted life years"—than comparative interventions.

These benefits are then weighed against the costs of different treatment strategies, with health systems using a value threshold—the upward limits of what they are willing to pay for a unit of health—to determine whether they will approve funding for a new treatment.[3]

To further quantify the value of their therapies, manufacturers also attempt to calculate the prevention value of their medicines: the amount of money saved and economic value created by preventing downstream disease (Maldonado Castañeda 2016). Such a view can be found in the marketing materials produced by the industry. In a fact sheet produced by the US pharmaceutical lobbying group PhRMA they argue that "every additional dollar spent on medicines or adherent patients with congestive heart failure, high blood pressure, diabetes and high cholesterol generated $3 to $10 in savings on emergency room visits and in-patient hospitalizations" and that "a 10 percent decrease in the cancer death rate is worth roughly $4.4 trillion in economic value to current and future generations" (Zirkelbach 2015). Dollars spent today on medicines, goes the story, creates future economic value. These cost-effectiveness value and prevention value methodologies thus aggregate the benefit across populations, quantify the economic value of this future benefit, and then compare this against competing or prior standards of care.

While clinical studies and trials have rendered visible the health benefits for early and ongoing treatment, these economic methodologies have quantified future benefits and provided a rationale for their valuation and pricing. These epistemic practices from medicine and health economics are part of a larger narrative shift, in which different actors carry new subjectivities toward health. From the vantage of governments, for example, health through therapeutic intervention is recast from an expenditure to an investment (Bach and Pearson 2015; Van Nuys et al. 2015). High prices, in many cases exceeding $100,000 for a course of treatment, are represented as signifiers for the potentiality of health improvements for its citizens. Across many therapeutic areas, each price sets the floor for the next product's price, as better health outcomes are to be valued with greater remuneration (Bach 2015; Vernaz et al. 2016). To sustain population health improvements, then, the state must be willing to continue to pay higher prices for more valuable therapies.

Patients are also brought into this mode of value, where the realization of future health is in part dependent on an individual's willingness to consume therapies now (Dumit 2012a). Such therapeutic consumption is far more than is necessary to maintain current health; this kind of future-oriented health through therapeutic intervention, not a healthiness that obviates treatments, has become central to the accumulation strategies of pharmaceutical companies. Treatment now, and into the future, is an asset, with health redefined into an asset—one that can be measured through statistical probabilities and then translated into economic value through cost-effectiveness research and prevention modeling. Along with shifts in the political economy of biomedical innovation, in which patents become monetized assets for accumulation, this assetized health carries stark implications for the possibility of curative futures.

Illusory Cures? The Limits of Assetization

In this context of valuing health in terms of the future, curative therapies would appear to lack any rivals. Medicines which end the progression of disease, not simply mitigate it, can radically alter the outlook for a patient and for populations—a future, as one recent conference considered, "free from disease" (Research America 2017). Of course, such a utopian forecast is far from our grasp, and the notion of curative therapies is not a monolithic one, given that the impact of a cure depends on the nature of the disease being targeted. In the case of an infectious disease like hepatitis C, for example, curative therapies are eliminating the pathogen from the blood stream, and over time may eliminate the disease from the population altogether. In the case of diseases that arise less from external pathogens but from genetic (e.g., cystic fibrosis) or gene-environment (e.g., many cancers) etiologies, a curative therapy may halt disease in an individual patient, but the disease would recur in the population with new incidences. The point of considering cures here, however, is not to capture all the complexities of different pathologies and their curative "matches," but to better understand the dynamics and limits of assetization for materializing the kinds of therapies so often imagined in our conversations about the future of biomedicine.

In thinking about assetization alongside cures, two potential crises feature centrally: first, the generation of continual growth in accumulation through therapeutic consumption and second, the naturalization of surplus health (and the justification of increasing prices) against the finitude of public budgets. To these, I also add a third potential crisis: the tendency of logics and practices of assetization to appropriate—and in many cases, elide—the public interest and the role of public investments in the breakthroughs behind curative therapies.

First is the problem of continual growth in accumulation in a stock-market, shareholder-driven political economy of therapeutic development. As I described earlier, throughout the life of an intangible asset, from the laboratories of a venture capital-backed company to control by a publicly traded pharmaceutical business, its value is determined by the future earnings *growth* that ownership and control of an asset might derive. Yet curative therapies are assets that eliminate the very possibility of growth on which its value as an asset relies. In this financial context, therapies for chronic diseases that require patients to take medicines over a long duration are—in contrast to cures—the optimal vehicle for accumulation. As Dumit (2012b, 81) puts it, "In too much drug research, cures get in the way of repeat revenue." Better than cures, from the vantage of a publicly traded company, are treatments for chronic pathologies such as high cholesterol and Alzheimer's, which show no sign of abating in terms of prevalence and incidence. With mitigator treatments, companies can accrue the kind of recurring revenue and growing accumulation structurally mandated by shareholders.

The logic can be used to understand why vaccine development is eschewed by companies in favor of mitigator treatments. In a study of different revenue models, Kremer and Synder (2003) put this view forward plainly: "Vaccines are more likely to interfere with the spread of the disease than are drug treatments, thus reducing demand for the product." A parallel to this situation can be found with the case of hepatitis C, a rare instance in which a cure has been developed. Gilead, the manufacturer that brought the curative drug *sofosbuvir* to market in 2013, made over $45 billion in revenue in the first three years of sales—a major key in tripling the company's annual revenue in that time-period (Roy and King 2016). Yet after its share price reached a peak near $120 in 2015, Gilead's market value dropped by almost half by early 2017 (Nisen 2017). The reason: by curing hepatitis C and reducing the potential patient population, Gilead was diminishing its opportunity for future growth. In a 2017 research report for industry leaders, financial analysts at Goldman Sachs posed Gilead's predicament with hepatitis C as a cautionary tale for the future of genomic therapies. In asking "Is curing patients a sustainable business model?" Goldman Sachs suggested that the central proposition of genomic advances—producing cures—represented a threat to the search for sustained cash flow that companies sorely needed (Kim 2018). Cures, in other words, do not work as financially valuable assets.

From this configuration arises a second problem with assetization and the possibility of curative futures: even if such therapies are developed, the valuation practices involved in a future-oriented view of health (e.g., value-based pricing) aim to naturalize increasing prices, thereby creating challenges for access to care for patients and health systems. When breakthroughs do occur, as in the case of hepatitis C, they often lead to significant potential improvements for patient and population health. In attempting to commensurate this potential improvement with price, health economics practices such as cost-effectiveness and prevention value legitimate rising prices with each progression in therapeutic advance.

These practices, however, place patients and public health systems in a challenging situation. Governments want to allow for innovation while also assuring access to new treatments. By making the value assessments I described earlier, health systems signal to drug developers that they are willing to pay more for better therapies while also setting a relatively high ceiling on the extent to which companies can push their monopoly pricing power. Such valuation practices present both budgetary and moral challenges for health systems. Public officials here are encouraged to "think like investors," as Birch and Muniesa (this volume) put it, to account for how paying for a given therapy at a certain price *now* may optimize a return on investment in terms of savings and quality adjusted life years *later.* Such stewardship of resources is thought to be well within the remit of public servants and policy-makers, and in the case of value-based pricing, is cloaked in the positive, aspirational view that *health is indeed an asset*—one requiring our generous remuneration as a society.

Yet health systems typically budget with a near-term time horizon (one to three years), while this future-oriented asset value of health is quantified over ten to fifty years. Locating this "value" within the health system—in the form of savings and health improvements—is a thorny and perhaps almost impossible accounting task. As people live longer and require more medical care for other causes, the valuation practices of "surplus health" can actually be used to argue for an ever-increasing consumption of treatments over a duration of one's life—which negates the very notion of future savings, as more treatments amount to growing spending (Dumit

2012a; 2012b). Ultimately, health systems with finite budgets often end up rationing treatment as a way of controlling expenditures and limiting opportunity costs in other areas of health and social spending, thereby making fraught ethical decisions about who should get access to treatment (Kesselheim, Avorn, and Sarpatwari 2016). Returning to the case of hepatitis C, health systems across Europe and the US initially limited access to only those patients in the latest stage of disease due to the price of the curative medicines, thereby diminishing the much touted prevention value of early treatment (Canary et al. 2015; Gornall et al. 2016; Iyengar et al. 2016).

Finally, the third crisis is the manner in which assetization tends to appropriate or elide considerations of public interest and the multiple roles played by the state in biomedical innovation. Assets, via the ownership and control protections granted by the state, allow for a measure of value by a singular metric: the accumulation that may be accrued in the future by its owners. In attempting to justify this accumulation through value-based pricing, biopharmaceutical companies have adopted the discourse used by many governments of value-based health delivery for patients and populations (Reinhardt 2015, 2016). This quantification and narrative strategy not only runs into the challenges noted in the prior point above, but this conception of value renders invisible the fundamentally social character of *value creation* in innovation, in which multiple actors contribute to the knowledge production process—with the state often a pivotal actor at the riskiest, uncertain stages.

In the US, for example, the National Institutes of Health (NIH) provided over $804 billion toward fundamental and applied scientific research from 1938 until 2012, with a significant bulk since the doubling of its budget in the 1990s (Lazonick and Mazzucato 2013). This funding has been behind major advances in biomedicine, such as the advent of molecular biology that served as the basis for the biotechnology sector (Vallas et al. 2011). Through this financing, the state, far from crowding out private actors as is often argued, has *created markets* for investors (Mazzucato 2016). The NIH, via the US government's small business innovation and research program, also provides direct investment to many of the early-stage companies that attract venture capital (Keller and Block 2013). By supporting innovation across multiple stages of the process, the state has played a pivotal role in some of the most important therapeutic leaps.

But in a "winner takes all" game of innovation, the company owning the right asset at the latest stages can accumulate the most. By contrast, the state—often the first investor—rarely receives any direct reward for their risk-taking (except for a small number of cases in which they retain a license or royalty agreement). Though the taxes paid on this revenue are considered an indirect return to the public, even these returns are diminished. In a world where intangible assets like capital are internationally mobile, multinational pharmaceutical companies move their control over assets across borders to tax havens, thereby reducing their tax rate in their home country (Lazonick and Mazzucato 2013). One study estimated that by domiciling assets (e.g., patents on approved medicines) in the favored tax haven of Ireland, US pharmaceutical companies have paid a tax rate of only 6 percent on over $100 billion in profits over the previous decade (Houlder et al. 2014). Such tax avoidance, however, threatens the very revenues used to fund the public investments in science that underpin the possibility for curative therapies and pay for the health systems that will deliver them.

On the one hand, assetization, as both political-economic and narrative transformations, has been used to attract vast sums of capital for innovation and make visible the potential for future health improvements at the level of populations. Yet on the other hand, some of the core logics of assetization—continual accumulation, the commensuration of value with rising prices, and an appropriation and elision of public interest—each presents distinct problems for a future of cures.

Concluding Thought Experiment: Competing Pathways for Confronting Curative Prospects

Despite the forbidding picture I have just sketched, what if the *longue durée* investments in scientific and technological advance mean that therapeutic breakthroughs in the form of cures cannot be avoided? Though such a techno-utopia may never appear before us as a possibility, a brief contemplation of such a thought experiment allows me to offer a few final possibilities for the transformations that have come to define the assetization of biomedicine.

In one scenario, the political economies and epistemic practices of biomedicine in an era of assetization may not change in any drastic fashion, and the status quo would remain. The incentives for companies to produce chronic mitigator treatments over cures would remain strong. Even though curative therapies would be produced in a few instances, their valuation in terms of high "value-based" prices might mean that a "rationing model" would persist and expand. Health systems, clinicians, and patients would be involved in perpetual struggles to gain access to new breakthroughs, as the value of curative therapies—indicated by escalating prices—would pose vexing questions of who should get such therapies first (Kesselheim et al. 2016; Kolata 2017; Reinhardt 2015, 2016). Public companies producing such therapies would need other products capable of producing sizable growth over time or risk becoming disposable businesses—ones that are acquired or go out of business once a disease is eliminated.

In contrast to the rationing model, a "public prize model" would entail severing the assetization process into two separate markets of assets—a market for the research and development of assets, and another market for the sales and distribution by the owner of the asset. Prizes for successful therapies, funded by taxpayers for disease areas of public health concern, would create sizable reward incentives as well as competition among teams of drug developers (e.g., $10 billion for a cure for HIV/AIDS) with the state then licensing this knowledge asset to generic manufacturers which could then distribute the medicines near the cost of production. Such a prize model has been raised as a possibility for drastically lowering the prices of new medicines and would pose a paradigm shift in the ways drug companies currently operate (Baker 2008; Love and Hubbard 2009). Yet even as the prize model has gained traction in discussions centered on reimagining biomedical innovation, generating the political will and momentum for such prizes across therapeutic areas remains a challenge—and may be more likely used in disease areas defined by citizen and patient-led advocacy as well as in cases of public health emergencies (e.g., antibiotic resistant pathogens, epidemic disease).

Another scenario for curative therapies involves a "Netflix model" of payment, in which patients and health systems become *subscribers* of particular companies or therapies. Rather than make per-treatment payments, members could pay a subscription fee to gain access to curative therapies (Goldman 2018). Such an arrangement would have the advantage of providing pharmaceutical companies with ongoing revenue while also enabling a payer—such as a health system—to space out their budgetary expenses and potentially guarantee greater access to medicines. This idea is being tested jn the state of Louisiana over access to hepatitis C treatments, in which the state would gain access to the therapies for five years for their poorest patients in exchange for an annual fee to pharmaceutical companies (Sagonowky 2018). While this strategy would address certain challenges associated with curative therapies (i.e., creating upfront access for cash-strapped health systems along with recurring revenue for companies), questions over how to calculate the prices of subscriptions, length of payments, and the extent of access to a given asset would likely remain deeply contested.

In the frontiers of disease where the current model of assetization prevails, still another possibility lurks: the "mortgage model." Payment for new breakthroughs, in this case, would not be limited by a public health system's finite budgets (or even private insurers' expectations of profitability), but rather would be facilitated by access to loans akin to buying a home. Upon a down payment (by an insurance plan, government, or individual patient), patients could gain access to a medicine, with a monthly or annual installment used to pay the total cost with interest over a duration of years. Through what a group of financial engineers at MIT recently proposed as "healthcare loans," the cost of cures is spread or amortized over many years with diversified pools of such loans securitized as financial products that can attract further capital (Montazerhodjat et al. 2016). These engineers argue that like a mortgage used to buy a home rather than rent it, healthcare loans would "buy health" through cures rather than "rent health" through continual payment of chronic treatments (mitigators). This expansion of payment financing may spur drug developers to direct science toward more curative therapies, given the ability to gain continual accumulation via interest and installments for such therapies. Yet this mortgage model could represent a different form of rationing with potentially unfair consequences, with patients' access to medicines tied to their socioeconomic positioning to participate in such loan programs. Here the logics of finance—with the valuation of the future enabling capital accumulation over a duration of time—would come to colonize not just the epistemic valuation practices of payment (e.g., value-based pricing) but the very financing of those payments.

The mortgage model represents the assetization of biomedicine in its most comprehensive and quintessential arrangement, from the development of new therapies in asset-driven financial markets all the way into their payment via securitized loans. That this proposal could even be widely discussed (published in one of *Science*'s journals) illustrates the powerful place that the processes of assetization have come to take in biomedicine. Surely the scientific and biomedical mysteries will be difficult to surmount in the development of curative therapies; yet it may be the turn to assets in the social organization and valuation of innovation that may prove most vexing.

Notes

1. These expectations are shaped by any number of events, from milestones such as the start of human clinical trials, clinical trial results, to an acquisition of a competing asset. Dramatic developments such as a clinical trial success or failure or a run of acquisitions can lead to asset bubbles, with share prices rapidly escalating, or bubbles bursting, with market value plummeting.

2. In following the history of disease since the 1970s, Dumit finds the rise of "prediseases" as emblematic of this conception (i.e., "pre-diabetes" and "pre-hypertension") denoting no felt symptoms by definition, but where risk factors could provide a rationale for treatment.

3. The use of this approach is common across Europe and gaining currency within the US, with the most prominent example the UK's National Health Service and its National Institutes of Clinical Excellence (NICE). In assessing new health technologies, agencies like NICE also set a "value threshold"—the upward limits of what they are willing to pay for a unit of improvement in health (measured in "quality adjusted life years"). See Claxton et al. (2008) and Bach (2015) for more on value-based assessments by health systems.

References

Abraham, J. 2010. Pharmaceuticalization of Society in Context: Theoretical, Empirical and Health Dimensions. *Sociology* 44 (4): 603–622.

Andersson, T., Gleadle, P., Colin Haslam, C., and Nick Tsitsianis, N. 2010. Bio-Pharma: A Financialized Business Model. *Critical Perspectives on Accounting* 21 (7): 631–641.

Bach, P. B. 2015. New Math on Drug Cost-Effectiveness. *New England Journal of Medicine* 373 (19): 1797–1799.

Bach, P. B., and Pearson, S. D. 2015. Payer and Policy Maker Steps to Support Value-Based Pricing for Drugs. *JAMA* 314 (23): 2503–2504.

Baker, D. 2008. The Benefits and Savings from Publicly Funded Clinical Trials of Prescription Drugs. *International Journal of Health Services* 38 (4): 731–750.

Berman, E. P. 2012. *Creating the Market University*. Princeton, NJ: Princeton University Press.

Biagioli, M. 2006. Patent Republic: Representing Inventions, Constructing Rights and Authors. *Social Research* 73 (4): 1129–1172.

Birch, K. 2017. Rethinking Value in the Bio-Economy: Finance, Assetization, and the Management of Value. *Science, Technology and Human Values* 42 (3): 460–490.

Block, F., and Keller, M. R. 2009. Where Do Innovations Come From? Transformations in the US Economy, 1970–2006. *Socio-Economic Review* 7 (3): 459–483.

Canary, L. A., Klevens, R. M., and Holmberg, S. D. 2015. Limited Access to New Hepatitis C Virus Treatment Under State Medicaid Programs. *Annals of Internal Medicine* 163 (3): 226–228.

Chiapello, E. 2015. Financialisation of Valuation. *Human Studies* 38 (1): 13–35.

Claxton, K., Briggs, A., Buxton, M., Culyer, A., McCabe, C., Walker, S., and Sculpher, M. 2008. Value Based Pricing for NHS Drugs: An Opportunity Not to Be Missed? *BMJ* 336 (7638): 251–254.

Csordas, T. 1994. *Embodiment and Experience*. Cambridge: Cambridge University Press.

Damodaran, A. 2017. Cost of Capital by Sector (US). *NYU Stern*, http://pages.stern.nyu.edu/~adamodar/New_Home_Page/datafile/wacc.htm.

DiMasi, J. A., Grabowski, H. G., and Hansen, R. W. 2016. Innovation in the Pharmaceutical Industry: New Estimates of R&D Costs. *Journal of Health Economics* 47: 20–33.

DiMasi, J. A., Hansen, R. W., Grabowski, H. G., and Lasagna, L. 1991. Cost of Innovation in the Pharmaceutical Industry. *Journal of Health Economics* 10 (2): 107–142.

Dumit, J. 2012a. *Drugs for Life*. Durham, NC: Duke University Press.

Dumit, J. 2012b. Prescription Maximisation and the Accumulation of Surplus Health. In *Lively Capital*, ed. K. Sunder Rajan, 45–92. Durham, NC: Duke University Press.

Fama, E., and Jensen, M. C. 1983. Separation of Ownership and Control. *Journal of Law and Economics* 26 (2): 301–325.

Gagnon, M.-A. 2016. Capital, Power and Knowledge According to Thorstein Veblen: Reinterpreting the Knowledge-Based Economy. *Journal of Economic Issues* 41 (2): 593–600.

Goldman, D. 2018. When We Find a Cure, Price It like Netflix to Ensure Access. *The Hill*, https://thehill.com/blogs/congress-blog/healthcare/401504-when-we-find-a-cure-price-it-like-netflix-to-ensure-access.

Gompers, P. A. 1994. The Rise and Fall of Venture Capital. *Business and Economic History* 23 (2): 1–24.

Gompers, P. A., and Lerner, J. 2004. *The Venture Capital Cycle*. Cambridge, MA: MIT Press.

Gornall, J., Hoey, A., and Ozieranski, P. 2016. A Pill Too Hard to Swallow: How the NHS Is Limiting Access to High Priced Drugs. *BMJ* 354: i4117.

Gregson, N., Sparrowhawk, K., Mauskopf, J., and Paul, J. 2005. A Guide to Drug Discovery: Pricing Medicines: Theory and Practice, Challenges and Opportunities. *Nature Reviews Drug Discovery* 4 (2): 121–130.

Harrison, C. 2011. The Patent Cliff Steepens. *Nature Reviews Drug Discovery* 10 (1): 12–13.

Helmreich, S. 2008. Species of Biocapital. *Science as Culture* 17 (4): 463–478.

Hopkins, M., Crane, P. A., Nightingale, P., and Baden-Fuller, C. 2013. Buying Big into Biotech: Scale, Financing, and the Industrial Dynamics of UK Biotech, 1980–2009. *Industrial and Corporate Change* 22 (4): 903–952.

Houlder, V., Boland, V., and Politi, J. 2014. Tax Avoidance: The Irish Inversion. *Financial Times*, https://www.ft.com/content/d9b4fd34-ca3f-11e3-8a31-00144feabdc0.

Hudson, K. L., and Collins, F. S. 2017. The 21st Century Cures Act—a View from the NIH. *New England Journal of Medicine* 376 (2): 111–113.

Iyengar, S., Tay-Teo, K., Vogler, S., Beyer, P., Wiktor, S., de Joncheere, K., and Hill, S. 2016. Prices, Costs, and Affordability of New Medicines for Hepatitis C in 30 Countries: An Economic Analysis. *PLoS Medicine* 13 (5): e1002032.

Jensen, M. C., and Meckling, W. H. 1976. Theory of the Firm: Managerial Behavior, Agency Costs and Ownership Structure. *Journal of Financial Economics* 3 (4): 305–365.

Keller, M. R., and Block, F. 2013. Explaining the Transformation in the US Innovation System: The Impact of a Small Government Program. *Socio-Economic Review* 11 (4): 629–656.

Kenney, M., and Patton, D. 2009. Reconsidering the Bayh-Dole Act and the Current University Invention Ownership Model. *Research Policy* 38 (9): 1407–1422.

Keshavjee, S. 2014. *Blind Spot: How Neoliberalism Infiltrated Global Health*. Berkeley, CA: University of California Press.

Kesselheim, A. S., Avorn, J., and Sarpatwari, A. 2016. The High Cost of Prescription Drugs in the United States: Origins and Prospects for Reform. *JAMA* 316 (8): 858–871.

Kim, T. 2018. Goldman Sachs Asks in Biotech Research Report: 'Is Curing Patients a Sustainable Business Model?' *CNBC*, https://www.cnbc.com/2018/04/11/goldman-asks-is-curing-patients-a-sustainable-business-model.html.

Kolata, G. 2017. New Gene Therapy Treatments Will Carry Whopping Price Tags. *New York Times*, September 11, 2017, https://www.nytimes.com/2017/09/11/health/cost-gene-therapy-drugs.html?mcubz=0.

Kremer, M., and Snyder, C. M. 2003. Why Are Drugs More Profitable Than Vaccines? *National Bureau of Economic Research*, http://www.nber.org/papers/w9833.pdf.

Lazonick, W. 2015. Share Buybacks: From Retain and Reinvest to Downsize and Distribute. *Brookings Institute*, http://www.brookings.edu/research/papers/2015/04/17-stock-buybacks-lazonick.

Lazonick, W., and Mazzucato, M. 2013. The Risk-Reward Nexus in the Innovation-Inequality Relationship: Who Takes the Risks? Who Gets the Rewards? *Industrial and Corporate Change* 22 (4): 1093–1128.

Lazonick, W., and Tulum, Ö. 2011. US Biopharmaceutical Finance and the Sustainability of the Biotech Business Model. *Research Policy* 40 (9): 1170–1187.

Light, D. W., and Warburton, R. 2011. Demythologizing the High Costs of Pharmaceutical Research. *BioSocieties* 6 (1): 34–50.

Love, J., and Hubbard, T. 2009. Prizes for Innovation of New Medicines and Vaccines. *Annals of Health Law* 18 (2): 155–186.

Lowy, D. R., and Collins, F. S. 2016. Aiming High—Changing the Trajectory for Cancer. *New England Journal of Medicine* 374 (20): 1901–1904.

Maldonado Castañeda, O. J. 2016. Price-Effectiveness: Pharmacoeconomics, Value and the Right Price for HPV Vaccines. *Journal of Cultural Economy* 10 (2): 163–177.

Mazzucato, M. 2016. From Market Fixing to Market-Creating: A New Framework for Innovation Policy. *Industry and Innovation* 23 (2): 140–156.

Montazerhodjat, V., Weinstock, D. M., and Lo, A. W. 2016. Buying Cures Versus Renting Health: Financing Health Care with Consumer Loans. *Science Translational Medicine* 8 (327): 327ps6–327ps6.

Mowery, D. C., and Sampat, B. N. 2004. The Bayh-Dole Act of 1980 and University–Industry Technology Transfer: A Model for Other OECD Governments? *Journal of Technology Transfer* 30 (1–2): 115–127.

Nightingale, P., and Martin, P. 2004. The Myth of the Biotech Revolution. *Trends in Biotechnology* 22 (11): 564–569.

Nik-Khah, E. 2014. Neoliberal Pharmaceutical Science and the Chicago School of Economics. *Social Studies of Science* 44 (4): 489–517.

Nisen, M. 2017. Gilead Mismanaged Its Gold Mine. *Bloomberg*, https://www.bloomberg.com/gadfly/articles/2017-02-08/gilead-earnings-mismanaging-a-goldmine.

Nitzan, J., and Bichler, S. 2009. *Capital as Power: A Study of Order and Creorder*. Philadelphia: Routledge.

Pellegrino, E. D. 1999. The Commodification of Medical and Health Care: The Moral Consequences of a Paradigm Shift From a Professional to a Market Ethic. *Journal of Medicine and Philosophy* 24 (3): 243–266.

Pisano, G. 2006. *Science Business*. Cambridge, MA: Harvard Business Press.

Rai, A. K., and Eisenberg, R. S. 2003. Bayh-Dole Reform and the Progress of Biomedicine. *Law and Contemporary Problems* 66 (1): 289–314.

Reinhardt, U. 2015. Probing Our Moral Values in Health Care: The Pricing of Specialty Drugs. *JAMA* 314 (10): 981–982.

Reinhardt, U. 2016. Value Creation and Value Shifting in Health Care. *Health Affairs*, https://www.healthaffairs.org/do/10.1377/hblog20160601.055099/full/.

Research America. 2017. Is a 'Disease Free' World Within Reach? *Research America*, http://www.researchamerica.org/news-events/events/national-health-research-forum/disease-free-world-within-reach.

Robbins-Roth, C. 2001. *From Alchemy to IPO: The Business of Biotechnology*. New York: Basic Books.

Rose, G. 2008. *The Strategy of Preventive Medicine*. Oxford: Oxford University Press.

Rose, N. 2007. Molecular Biopolitics, Somatic Ethics and the Spirit of Biocapital. *Social Theory & Health* 5 (1): 3–29.

Roy, V., and King, L. 2016. Betting on Hepatitis C: How Financial Speculation in Drug Development Influences Access to Medicines. *BMJ* 354: i3718.

Sagonowky, E. 2018. Louisiana Seeking Comments on Netflix Model for Hepatitis C Drugs. FiercePharma, https://www.fiercepharma.com/pharma/louisiana-seeking-comments-netflix-model-for-hep-c-drugs.

Sample, I. 2016. Cure All Diseases? Zuckerberg's Plan Is Brilliantly Bold. *Guardian*, https://www.theguardian.com/commentisfree/2016/sep/23/priscilla-chan-mark-zuckerberg-cure-diseases-facebook-initiative-brilliantly-bold.

Scheper-Hughes, N., and Lock, M. 1987. The Mindful Body: A Prolegomenon to Future Work in Medical Anthropology. *Medical Anthropology Quarterly* 1 (1): 6–41.

Scherer, F. M. 2004. The Pharmaceutical Industry—Prices and Progress. *New England Journal of Medicine* 351 (9): 927–932.

Sharp, L. A. 2000. The Commodification of the Body and Its Parts. *Annual Review of Anthropology* 29 (1): 287–328.

Slaughter, S., and Rhoades, G. 1996. The Emergence of a Competitiveness Research and Development Policy Coalition and the Commercialization of Academic Science and Technology. *Science, Technology, & Human Values* 21 (3): 303–339.

Sunder Rajan, K. 2006. *Biocapital*. Durham, NC: Duke University Press.

Sunder Rajan, K. 2012. Pharmaceutical Crises and Questions of Value: Terrains and Logics of Global Therapeutic Politics. *South Atlantic Quarterly* 111 (2): 321–346.

Taussig, M. T. 2010. *The Devil and Commodity Fetishism in South America*. Chapel Hill, NC: University of North Carolina Press.

Vallas, P., Kleinman, D. L., and Biscotti, D. 2011. Political Structures and the Making of U.S. Biotechnology. In *State of Innovation The U.S. Governments Role in Technology Development*, ed. F. Block and M. R. Keller. Boulder, CO: Paradigm Publishers.

Van der Zwan, N. 2014. Making Sense of Financialization. *Socio-Economic Review* 12 (1): 99–129.

Van Nuys, K., Brookmeyer, R., Chou, J. W., Dreyfus, D., Dieterich, D., and Goldman, D. 2015. Broad Hepatitis C Treatment Scenarios Return Substantial Health Gains, but Capacity Is a Concern. *Health Affairs* 34 (10): 1666–1674.

Venter, J. C., and Cohen, D. 2014. The Century of Biology. *New Perspectives Quarterly* 31 (1): 28–37.

Vernaz, N., Girardin, F., Goossens, N., Brügger, U., Riguzzi, M., Perrier, A., and Negro, F. 2016. Drug Pricing Evolution in Hepatitis C. *PLoS ONE* 11 (6): e0157098.

Weinstein, M. C., Torrance, G., and McGuire, A. 2009. QALYs: The Basics. *Value in Health* 12: S5–S9.

Zirkelbach, R. 2015. Five Essential Truths about Prescription Drug Spending. *PhRMA*, http://catalyst.phrma.org/the-five-essential-truths-about-prescription-drug-spending.

II Turning Infrastructure into Assets

5 High-Speed Contradictions: Spanish Railways between Economic Criticism and Political Defense

Natalia Buier

Introduction

The last few years have seen increased attention devoted to the rise of infrastructure as an asset class as analysts have begun to comment on the changing role of the private sector in infrastructure development in the aftermath of the global financial crisis (e.g., Haughton and McManus 2012; Hildyard 2012; O'Brien and Pike 2015; O'Neill 2009). The problem of infrastructure as an asset class is typically understood as a phenomenon of capital opening new avenues for the creation and monetization of streams of revenue, or "assetization" (Birch and Muniesa, this volume). Typically, the problem of infrastructure as an asset class is studied at the level of infrastructure finance, as a transformation of the relationship between the private and the public sector, where the latter is shown to be creating new opportunities for the extraction and accumulation of profit primarily through the design of new investment vehicles (Hildyard 2012; O'Brien and Pike 2015).

The rise of private equity funds and private-public partnerships has been seen as the defining attribute of the process of conversion of infrastructure into assets. The analysis of this process is explicitly or implicitly underwritten by a shared view about the transformation of the role of the state in the provision of infrastructure. The provision of infrastructure, it is widely believed, until recently has been a case of state provision of public goods. This is a process that can be best explained by the specific properties of infrastructure, which usually requires outlays of capital that cannot be secured without state intervention.

Furthermore, the life cycle of infrastructure is typically one that makes the expected time frame of profitable returns unattractive to private capital. Recently, however, there has been an increasing presence of private finance in infrastructure development. This is primarily seen as the result of the fiscal crisis and budgetary restrictions affecting national states, which have seen their ability to invest in infrastructure significantly reduced. The strain on national budgets does not alone explain the increased presence of the private sector in infrastructure development. This requires not only the weakening of the ability of the state to invest but also the existence of what appear to be profitable investment opportunities. The private sector has been able to successfully promote the conversion of infrastructure into a new asset class by assembling new investment vehicles that make possible the sidestepping of typical risks associated with investment into infrastructure (Hildyard 2012; O'Brien and Pike 2015).

In what follows I present the case of an infrastructural mega-project that does not conform to this story but rather defies most expectations about the contemporary conversion of infrastructure into an asset class.[1] Yet, I argue, this is fundamentally a case of the conversion of infrastructure into a profitable stream of revenue as well as one about the failures and resistances that surround it (see also Braun, this volume). Rather than simply being an outlier case in an otherwise existing trend, the project I analyze—Spanish high-speed rail (HSR)—alerts us to the possibility of a more complicated dynamic when it comes to turning infrastructure development into lucrative business opportunities. The creation of profitable investment opportunities in the field of infrastructure cannot be understood just as a case of transition from publicly provided goods to privately financed projects. Rather, it represents a process of the reconstruction of the public sector in line with market criteria and the imperatives of capital accumulation.

In order to show the ways in which HSR is embedded in the reconstruction of the contemporary railways as a profit-driven enterprise, I start with a broader description of the contemporary Spanish HSR project. From this I move on to the origins of the project and discuss the way in which the project emerges as part of a broader shift to a commercial railway, the origins of which are found in the 1980s. From discussing the origins of the project I turn to the contemporary debates that surround HSR. The dynamic of contestation and defense of HSR reveals the way in which the failure to construct HSR as a profitable asset does not actually result in a broad contestation of the market model of infrastructure provision. Within dominant discourses both the critique and defense of the HSR project represent different approaches to defending the "market order of worth" (Davies 2013, 37).

The confrontation between the proponents and critics of the project relies on the continued production of factual evidence that allows for the quantification of the results of HSR, a process that I describe as the "number wars." As an alternative to dominant discourses both in favor and against Spanish HSR, I briefly introduce the anticapitalist critique of it. This thread of contestation, although not fully autonomous since it depends on some of the factual repertoire of the dominant critical discourse, points to the way in which both the economic criticism and the political defense of HSR can be rejected by challenging the identity between the state and the public.

Alta Velocidad Española

The importance of Spanish HSR (*Alta Velocidad Española*, or AVE) among European infrastructure projects is immediately signaled by its magnitude: hailed as the most important infrastructure project in Spanish history, it has resulted in what is today the longest HSR network in Europe and the second largest globally (for recent comparative data, see European Court of Auditors 2018). Far less known than the Japanese or French precedents, it has been the priority of Spanish infrastructural development programs for almost two decades. Its origins go back to the mid-1980s when the second socialist government of Felipe González took the historic decision to construct the first HSR line on the Madrid-Seville route. Inaugurated in 1992, a year of seminal importance in the recent history of Spain, it became a key symbol of an expansive, modernizing Spain on its route to full European integration. The properties of today's network were definitively established in the 2000s, when two successive national infrastructure plans maintained the objective of connecting Madrid to all the regional capitals by HSR, thus putting Spain on track to becoming the leading European developer of HSR.

Rather than a retreat of the state from infrastructure finance, this is a project that has almost entirely relied on public funding. While defenders of the project highlight the role that European development funds have played in its development, public borrowing, at both the national and the regional level, has been the essential financial instrument of a project that has drawn in resources on an unprecedented scale (Audikana 2015; European Court of Auditors 2018). This, put briefly, means that the state has remained the key actor in assembling the resources and institutions that have made possible the development and management of HSR.

Before moving on to a more detailed discussion about the origins of the AVE, a few remarks about the structural properties of HSR are necessary. Much of the official discourse that surrounds HSR, both in Spain and at the EU level, is focused on HSR being an environmentally friendly transport infrastructure. The official Spanish discourse, at both national and regional level, also insists on the role of HSR in promoting territorial cohesion. The underlying assumption in this chapter is that both these discursive directions obscure the fundamental reality of HSR as a greenwashing instrument and as a disarticulating infrastructure. My focus here, however, is not on building the critical case against the AVE but on providing an alternative vision to dominant narratives about infrastructure as an asset class and on showing how disputes around the success or failure of HSR as a profitable infrastructure reveal the need for a continuous production and reproduction of the ideological foundations of the project of a commercial railway. These debates show us the deeper ideological conversions required for the successful transformation of an infrastructure into an asset class.

HSR is, as a transport infrastructure system, a key element in the organization of territory. The "spatial order" of HSR favors central urban nodes and end destinations, at the expense of intermediate regions (Ureña 2012). The underbelly of the discourse of cohesion, at a national or European level, is the widespread experience of disconnection that is cosubstantial to the development of HSR. For Spain, a country which from the 1950s onward experienced an accelerated growth of the urban regions, with the 1960s and 1970s processes of concentration in metropolitan areas, and the more recent dynamics of peri-urbanization of the decades of the 1980s and 1990s (Ureña 2012, 79), this has meant that HSR inserted itself into a rather straightforward dynamic: between 1991 and 2007 "the part of Spain that was growing did so progressively and the Spain that was in decline accentuated its regression" (Ureña 2012, 64). Despite the efforts of proponents of HSR to prove the potential benefits of HSR in terms of the economic growth of smaller municipalities on the network, the bulk of the evidence is against this belief, increasingly seen as an HSR myth. Regional disparities are even more striking in the case of Spain, where the radial network of HSR and the strengthening of Madrid as central node lead to significant inequalities in the distribution of benefits within the network.

The number wars for and against HSR often reach significant levels of seeming methodological sophistication, although it is becoming increasingly apparent that proponents of HSR need to turn to different legitimation grounds as critics seem to be gaining the upper hand. The disputes often cloud the basic realities of HSR, which are much better expressed and captured by the daily experience of users having to turn to travel by bus as conventional rail services get canceled, or by figures that leave little room for doubt. Conventional rail makes it possible to have stations every 15 to 30 km; the technological properties of HSR typically require stations to be separated by distances over 150 to 200 km. Straightforward evidence of the so-called "tunnel effect" of HSR is perhaps better explained by the images of desertification that those negatively affected by the development of the AVE invoke. Where academics and experts see tunnels connecting end points, those who confront the experience of the AVE from behind the fences separating the expensive infrastructure speak of deserts. Nonetheless, an ethnographic turn to the number wars can enhance our understanding of the AVE as a historically specific configuration. If the meaning of the most ambitious infrastructural development program in the history of Spain is certainly not exhausted in winning the number wars, understanding how to read them is a key element in revealing alternatives to the existing order.

Origins of the AVE

Approved in 1986, the first HSR line in Spain, Madrid-Seville, was hailed as a project that would finally halt the trajectory of decline that had characterized the national railways during the previous decades. Briefly, the history of railways during Francoism was one of progressive marginalization. The first two decades of the Francoist railways were marked by the legacy of the civil war. The physical infrastructure, heavily damaged and already antiquated, was overseen by a complex bureaucratic hierarchy and maintained by what was considered an oversized workforce. Beginning in the 1960s, the railways were targeted by a series of reforms that were primarily aimed at modernizing the infrastructure and the rolling stock. The modernization programs already addressed the management of the railways as well, and the reduction of the workforce became a priority. While important transformations were carried out during the 1960s and the early 1970s, these did not succeed in containing the overall direction of decline. This was manifested in two simultaneous trends: the rising deficit of the national railway company and a diminishing market quota for railway transport (Muñoz

Rubio 1995). During the first years of the democratic period RENFE, the national railway company, became a powerful symbol of the legacy of Francoism and the inefficiency of public management. It was trapped in the contradictions generated by, on the one hand, its importance as a key public company, and on the other hand, the increasing marginality of rail as a mode of transportation. The massive deficits that the company ran up came to stand in as shorthand for both, as the railways became widely seen as an inefficient, antiquated institution, an exponent of the old regime, and the bastion of a privileged workforce (Comín et al. 1998; Muñoz Rubio 1995).

The first years of post-Francoism saw the development of sectorial plans that aimed to modernize the railways. The first post-Francoist national railway plan, the PGF (*Plan General Ferroviario-General Railway Plan*), developed under the transitional government of UCD (*Unión de Centro Democrático-Union of the Democratic Centre*), explicitly addressed this reality and proposed a massive investment plan that would aim to correct chronic underinvestment in the railways, modernize the infrastructure, and expand the workforce, with the goal of providing a reliable service that would restore the railways to their former glory. It is difficult to decide, retrospectively, whether this first investment plan was a well-crafted diversion that would forestall the possibility of unrest in what was at the time the largest public company, or whether it represented the temporary victory of a faction of railway management and public administrators that still believed in the possibility of restoring conventional rail to its former centrality. It is certain that this plan echoed the concerns that emerged from the oil crisis, which had briefly managed to open a crack in the ideological hegemony of highway transport, if it never significantly altered its centrality in infrastructure policy. What can be known from secondary sources is that this short-lived post-Francoist modernization plan was enthusiastically embraced at the level of the company. Its abandonment marked the opening of a radical shift in railway policy (for a broader discussion, see Buier 2016).

The decision to abandon the PGF was taken by the first socialist government of the democratic years and in its aftermath a commission for the study of the situation of the railways was established. Commonly known as the Roa Commission, because of its president, Carlos Roa, its work sealed the death of the PGF as railway policy, establishing deficit control and profitability as the objectives of RENFE (*Red Nacional de los Ferrocarriles Españoles*). The early work of the Roa Commission was instrumental in establishing a shift in how the deficit of the company was viewed. If previously the deficit was seen as economic in origin, with underinvestment as a primary cause, a new consensus was put in place in the 1980s, as arguments about the managerial origin of the deficit gained weight. Company reorganization thus became essential to the pursuit of economic profitability. This is how Gonzalo Martín Baranda, socialist railway manager and author of an autobiography, remembers the period: "In order to close lines the cost of the train for the citizens had to be exposed to public opinion. This generated in the people an animosity against the *ferroviario* that was lived through with anger in RENFE" (2011, 68).

During the first socialist government the biggest closure of railway lines took place. The 1980s also resulted in a drastic reduction of the workforce, as a new philosophy of human resource management was put in place. In the words of Gonzalo Martín Baranda:

> It was that team, the first one that estimated and compared the costs of accidents, pollution, the time lost between the highway and the railway. I usually give a phrase by Paracelsus which says: "The only things known are those that can be counted and measured." (Martín Baranda 2011, 71)

The computation of social costs and the task of rationalizing management were enthusiastically pursued and aided the objective of revealing the way in which the previous generations of managers had sidestepped the objectives of economic profitability. The entire architecture of the company had to be changed in order to reflect and aid the public company in the effort to emulate the successful recipes of the private sector. The autonomy of the company, a tenet of promarket policies and a long-standing contentious issue for the railways, became central to the dominant managerial vision of the 1980s and found a strong continuity in the presidency of Mercè Sala, the first woman to be the president of the company, also a socialist appointee. Internal reorganization on the basis of private sector imperatives was one route toward achieving a competitive railway.

The origins of the AVE are firmly rooted in this context. Technological modernization was the twin process of internal reorganization. If internal reorganization would secure the closing of the gap between the railway company and other actors in the broader entrepreneurial landscape, technological modernization was the route to modal specialization, or the niche in which it was believed railways could compete with alternative means of transport. The AVE was born in a context of intense debates about the competitive specialization of the railways. Effectively, the demise of the railways as a hegemonic mode of transport found its definitive legal expression in the second half of the 1980s. With the LOTT (*Ley de Ordenación del Transporte Terrestre*), it was finally established that the paradigm of the railways as a privileged transportation monopoly should be abandoned in favor of a transport market where each mode of transport specialized according to its competitive advantages. At the time it was firmly believed that HSR long-distance passenger services would prove competitive in relation to air transportation.

This was also the time when the first efforts to separate the balance sheets for railway operations and infrastructure were made. European policy of railway liberalization relies on vertical unbundling which initially took the form of the separation of the financial results of infrastructure management and service provision. However, the first efforts to go in this direction occurred before significant transformations in transport policy at the EU level and are tied to the early days of the AVE.

If the major early investments required by the development of HSR were the target of some criticism, the success of the Madrid-Seville line managed to support the idea that under well-managed conditions HSR operations could be profitable. In the context of the changes in EU transportation policy in the 1990s, the dynamics set in motion already in the 1980s in Spain could become firmly established. The demands of restructuring inscribed into the European legislation are aimed at separating profitable services from the so-called public services which can be supported through public subsidies. Yet, RENFE and its subsequent divisions never managed to meet the deficit targets set at the national and European level. Repeated write-offs of debt, company restructuring, and various forms of financial engineering have been mobilized throughout the years in an attempt to control the deficit or produce the appearance of a company registering profits. The development of the AVE in particular was a challenge in this respect, given it required the concentration of resources on an unprecedented scale.

The AVE as an Object of Debate

Thirty years later, the AVE has been firmly established in the converging infrastructure policy of the two main Spanish parties, PP (*Partido Popular*) and PSOE (*Partido Socialista Obrero Español*). Nor was the development of HSR significantly affected by the unfolding financial crisis. At the same time, though, criticism of the AVE has gained momentum and the defense of the project has become ever more entrenched in political debates about territorial cohesion and solidarity. The most visible criticism is an economic one. This sees the entire project as an irrational squandering of resources with the promise of unproven returns. A salient incarnation of this line of opposition can be found in the cost-benefit analysis (CBA) of the AVE. In recent years, there has been a multiplication of studies that look at Spanish HSR through this lens (Albalate and Bel 2012, 2011; Bel 2007, 2010a, 2012; De Rus and Nombela 2007; De Rus and Roman 2006; Mendez et al. 2009). Taken up primarily by (transport) economists, it typically involves the analysis of questions of profitability, demand, regional economic impact, and environmental benefits of HSR.

If the proponents of this type of analysis usually like to maintain the appearance of a balanced tone, highlighting that under very special circumstances HSR might prove to be a justifiable investment, the practical conclusions most of the time lead to an unambiguous rejection of this infrastructure. The special conditions that HSR must meet are primarily related to estimated demand on a potential new line and expected returns on operation. This, the argument usually goes, only makes HSR worthwhile in the situation where it meets the function of alleviating congestion on corridors linking densely populated metropolitan areas. The verdict is out on this, we are told, with HSR so far only proving profitable in two cases: the Tokyo-Osaka line and the Paris-Lyon one. No other HSR project to date has proven economically profitable. This is backed up by evidence that shows that far from being able to recover the cost of investment in the foreseeable future, the AVE also generates losses at the operating level.

Regional economic development is an idea well entrenched in the repertoire of the defenders of the AVE. The pro-HSR lobby on the regional level has essentially relied on the argument that it brings prosperity in the construction phase as well as in the operational one, by integrating towns into the most advanced transport network in Europe. To be left outside the network consequently became a symbol of being cut off not only from prosperity but from claims to Europeanness itself. CBA advocates, however, are profoundly skeptical of this argument too. The counterargument is convincing, as critics point out that there is no conclusive evidence about the growth of smaller towns following the arrival of the AVE. Even in the cases where growth has been observed, there is not sufficient evidence to attribute it to the AVE. Finally, according to CBA, the environmental record is also much more complicated than defenders would have us believe. If the AVE is clearly more environmentally friendly than air transport, its relative position in relation to transportation by bus and car is not clear. Even the latter, with a certain level of occupancy, might prove to be more energy efficient. In relationship to the environmental record, CBA usually highlights that efficiency calculations for the environmental impact of HSR do not take into account the major impact of building the new infrastructure, focusing simply on infrastructure in use.

The Spanish proponents of CBA bring to the problem of financing public works and infrastructure a range of calculative techniques which they reify as a fixed method for comparing costs to benefits. CBA, it is believed, can serve as a tool for disciplining a political process that is fundamentally irrational. Unsurprisingly, its proponents present it as offering an unfailing standard of rationality to the problem of state funded development. This line of critique of the development of the AVE presents itself in opposition to the dominant logic driving defenders of the project. However, upon closer scrutiny it is revealed that the advocates of CBA represent simply one incarnation of a form of "militant quantification" (Porter 1995, 187) grounded in the idea of economic estimation as neutral.

As Theodore Porter (1995) has shown, cost-benefit analysis became a "respectable economic specialty" in the United States in the 1950s. But what his analysis reveals is that the earlier history of standardization of CBA, intimately tied to the politics of quantification surrounding large public works, is far from a fixed method with uncontroversial application (see also Porter 1992). Proponents of CBA present it as a tool of universal validity, although it is better described as a provisional set of techniques for monetary valuation. These techniques represent a particular and historically specific answer to the question of what constitutes a benefit and what constitutes a cost. Proponents of CBA present defenders of the AVE as falling outside the realm of rational economic calculation. But behind this surface appearance of the conflict what we can see is the actually existing diversity of cost-benefit practices. Although governmental rhetoric is sometimes clad in nationalist overtones defending unquantifiable benefits, in practice the drivers of the project rely on their own preferred measures for quantification and valuation. This means that extensive use is made of quantifying intangible benefits and forever expanding the reach of monetary calculation to indirect benefits. Official planning documents and administrative practice favor the monetary expression of time savings. The official line of defense is also grounded in the relentless search to measure indirect benefits derived from the development of HSR, from job creation to increases in property prices and incentives for investment.

Where CBA proponents see a conflict between economic and political logic, between the rational expression of economic calculation and the irrationality of easily corrupted political planning, what is actually at work is the ongoing conflict between different ways of extending the realm of calculation and subordinating planning to the needs of the market. The main difference between those who employ CBA as a tool to oppose investment into HSR and those who defend investment into it is not given by different allegiances to the market. Rather, the difference resides in competing visions regarding the concrete workings of economic estimation. The use of CBA by its Spanish proponents is fully consistent with the analysis by Porter (1995). This is a paradigmatic case of the search for mechanical objectivity, or the attempt to establish a decision-making routine that, "once set in motion by appropriate value judgments on the part of those politically responsible and accountable, would—like the universe of the deists—run its course without further interference from the top" (Porter 1995, 189). CBA remains a planning tool firmly entrenched in the terrain of anti-democratic expert rule in the service of markets.

It is not surprising that proponents of CBA can represent support for the AVE as following a logic of political, rather than economic planning. During the Aznar and Zapatero governments the development of HSR has been inscribed and presented as an instrument of territorial cohesion and as a political choice. Public inaugurations of new lines have become a ritual display of regional development and European integration, as politicians across party lines claim patronage. Words such as the ones heard early on at the inauguration of an HSR line in Andalucía have been firmly settled as the common tropes of the festive inaugurations: Zapatero praised the development of Andalucía during the last three decades of "freedom and democracy." The region is, for the prime minister, a region that is "modern, transforming and growing at a pace above the Spanish average. It is firmly and decisively walking the path of full integration to Europe."[2]

Illustrative of both the convergence and the battle for symbolic patronage are incidents such as the fact that the absence of an important socialist official from the inauguration of a line can appear as sectarian and divisive. During the 2015 electoral campaign, the failure of prime minister Mariano Rajoy to invite his predecessor Zapatero to the official opening of a line begun during the latter's mandate could be seen as "the end of a tradition."[3]

During my fieldwork I have heard many times a certain form of subtracting oneself from an evaluation of HSR. Its generic representation would be: "I cannot tell you if the AVE is good or bad, this is a question of political will. The government must decide if they want to construct a new line or not, but this cannot be decided in economic terms, it cannot be formulated as an economic question." Here, then, was the same logic that animated the most radical critics of the project. "The railway deficit is a problem that cannot be solved as an economic problem, it has been made into one through political will," Daniel, an engine driver, had told me. "It should not be set out as an economic problem." "If you look at this as an economic problem it does not make sense. But it cannot be decided like this. It is a question to be decided at the governmental level, it must be decided whether this new line is wanted or not," Miguel, a SEMAF[4] unionist, had argued. But in his argument there was more than evasiveness and an encroaching understanding that the unfathomable investment figures for the AVE had started backfiring with talk of indebtedness. His argument echoed a managerial obsession that has haunted RENFE for decades.

In the long history of the disputes about the question of the autonomy of the public company, freedom from governmental intervention has implied several things. Prominently it has been used to highlight that such autonomy could allow for a rational management of economic resources, and that this way the functioning of the railways could be set firmly on a commercial basis. But the corollary of the argument has also been one that aimed to free the company from the investment decisions proper. RENFE would act, of course, as a modern company in the pursuit of commercial objectives, but it should not be an administration with the power to decide what lines should be built. This responsibility should belong to the government.

Still, upon closer scrutiny the so-called political defense of the AVE is revealed as firmly anchored in a broader commitment to quantification. In addition to the range of calculative techniques designed to maximize the monetary benefits of the project, the faith of HSR has also been intimately connected to the challenge of containing the financial deficit resulting from its development. Before I look at this in more detail, I discuss a form of opposition that aims to challenge the order of worth of the market.

Challenging the Separation of the Economic and the Political

CGT, *Confederación General del Trabajo*, is the confederation that represents the majoritarian sector of Spanish anarcho-syndicalism. The railway section of CGT is among the strongest in the confederation, and the 2015 elections, despite a frontal attack against the union aimed at reducing its representation, secured the presence of CGT in the works' council of both RENFE and ADIF (*Administrador de Infraestructuras Ferroviarias*, the Spanish railway infrastructure manager), with two members in each.

In the railway sector, CGT pushes for an alternative "public and social railway" (*ferrocarril público y social*). While the most recent articulation of the meaning of the proposal is found in a 2012 document, systematic treatments of this position could already be found in 2001 when the union published what they themselves consider to be the most complete document devoted to the railways.[5] Put succinctly, the CGT alternative can be summed up in ten demands, which, in turn, can be summarized as follows: the railways must continue to be a public service, placed above economic criteria favoring the interest of the few. The railways must be maintained as public property. Investments in railways must prioritize conventional rail, and safety must be guaranteed above all other criteria. The railway system must maintain the concept of integrated planning and services. The accounting criteria must take into consideration the savings in external costs. Users must benefit from these savings in the form of adequate service provision. Accessible and subsidized tariffs must support the development of railway service. Railway transportation must be promoted as a priority transportation service. And, finally, a common employment framework across the sector is needed in order to guarantee work conditions as well as safety, both in terms of work safety and transportation safety.

Flipping through CGT leaflets immediately alerts you to a story told differently. As opposed to the timid recuperation of state ownership that CCOO and UGT (the two majority union confederations) sketch, where the state-owned railway is at best opposed to the private one, the brief historical sketch that the CGT promotes for general audiences speaks of the cyclical history of the railways. Twentieth century railway history, we are told, is a history of oscillation between public and private ownership, where liberalization, privatization, and (re)nationalization represent different moments in processes of capital accumulation. There are phenomena that cut across the public/private divide, it is argued. The discourse of the inefficiency of public management is one of them; the other is the constant issue of the railway deficit, an ever-present pressuring instrument. The capitulation of railway management to economistic criteria occurs in both phases, with the state implementing policies that are designed to benefit capital and the private accumulation of profit. So while defending the public railway, CGT appears to qualify the history of public ownership as state ownership.

The case against the AVE that CGT builds can only be understood as an extension of the broader vision of the railway that the union promotes. The AVE is, in opposition to the public and social railway, an elite railway, built for the benefit of the few at the cost of the many. Subordinated to a model of territorial development that the union rejects, the AVE appears as an element in a broader infrastructural policy that has placed profit, at all costs, ahead of sustainability, broadly understood. One meaningful point of friction between the critique of the AVE as an extension of the defense of the public and social railway and the need to engage with the hegemonic framework is the union's reliance on the data generated by CBA. Although CGT is a staunch defender of a railway model that is placed above strict criteria of profitability, their daily work requires an engagement with hegemonic discourses. This, in practice, has meant that CGT has relied on the type of data provided by CBA to prove that the AVE represents, from the economic point of view, a failed model. The union's argumentative strategy oscillates between a double-edged critique with clear priorities and the ambiguities of resorting to the factual repertoire of liberal economics in order to defend a nonliberal railway model (for details see Buier 2016).

Superficial Contradictions and Ideological Convergence

Here we are then, with thirty years of AVE, facing a situation that appears rather paradoxical. The strongest line of critique of the AVE, the economic one, emphasizes the political criteria in infrastructure policy. A more reserved and apparently neutral positioning toward the AVE, such as in the case of some of the SEMAF unionists I interviewed, highlights the same divide between the political and the economic, but delegates decision to the political realm (echoing some of the arguments of 1980s New Public Management). The anticapitalist critique of the AVE, as seen in the discussion about CGT, also operates with the same political and economic distinction, but stresses the dominance of the economic over the political. On the other side, the defense of the AVE has come to increasingly be formulated in terms of territorial cohesion and regional solidarity. These arguments are advanced through a form of claim-making according to which the decision to build this new infrastructure cannot be decided simply on the basis of a certain type of economic evidence. At least on the surface, then, it would appear that a government fiercely committed to a politics of austerity and privatization is defending a certain sector from the encroachment of economic criteria. In the following section I look more closely at this apparent contradiction.

To accept the representation of the main arguments for and against the AVE as a clash between economic and political rationalities is erroneous. But so is the conclusion that opponents of the AVE have carried their arguments to similar conclusions. The way I reconstruct the arguments between defenders and opponents of the AVE is first of all meant to highlight the shared market orientation between the governmental defense of HSR and opposition to it as reflected in CBA. Differences between the two begin to emerge once the question of the relationship between the market and the government is articulated. As seen, critics of HSR who resort to CBA oppose the logic of the market to what they see as the centralizing and centralized planning at the level of the national government. The form of the argument is that of a "preference for governmental agnosticism as a form of liberal neutrality" (Davies and McGoey 2012, 77), which is why CBA does not consider itself prescriptive but merely claims to provide the empirical data for policy-makers. However, substantively, CBA is the empirical and methodological repertoire of a view that would fully entrust transport planning to the market.

In opposition to this, the defense of the AVE has taken the appearance of an argument for limiting the reach of the market. This, again, is misleading. When placed in their broader context, the arguments about territorial cohesion and solidarity reveal their meaning not as against the ethos of the market, but rather as a different defense of the "order of worth of the market" (Davies 2013). This is a view that ascribes a different role to the state, which maintains the role of organizing on the national scale the material basis for the successful operation of markets. In opposition to this, arguments such as those favored by CGT maintain that the railways should not be subjected simply to market-based forms of valuation. It is indeed complicated to always separate this uncompromised position from the tactical repertoire and the factual evidence it employs. CGT, just like the environmentalist critics of HSR, often relies on "social cost" calculations or the calculation of "externalities." This is evidence that is summoned in order to prove not only that the railways could prove competitive but also that the only reasons other modes of transport appear as competitive is because the hidden costs associated with them are not taken into account. And this repertoire of factual evidence remains the result of extending market calculations to areas that were previously considered nonmarket. But if this repertoire of calculation is summoned, this remains subordinated to the argument that profit seeking should not be the foundation on which transport planning and territorial development occur. And this is most clearly articulated in not only the rejection of HSR but in the defense of conventional rail.

Competing Calculative Devices

The reconstruction of the arguments between defenders and opponents of HSR could suggest that in effect the competition occurs on the terrain of disputing the limits of governmental intervention in markets and in particular transportation markets. But the actual unfolding of the conflict does not merely oppose different promarket visions, it involves the mobilization of an entire range of rival calculative devices. While it might appear that the EU budgetary cutbacks and fiscal consolidation are recent enemies of the development of HSR, to exceptionalize the current pressures is misleading. In effect, the railways and HSR development have been facing fiscal policy constraints throughout their entire existence in the post-Francoist period. The implementation of HSR occurred around the time of Spain's accession to the EC in 1986. The plans for the massive extension of the network were developed as Spain was preparing for the adoption of the euro. Finally, during the recent crisis HSR has remained a privileged infrastructure project. These are also periods during which the question of public deficit management was paramount to fiscal policy in Spain. This is not to say that during the periods considered to be economic crises the overall rhythm of infrastructure development was not affected, since in effect the commercialization agenda of PSOE during the 1980s and the resulting consequences for the railway company were put in place in the early 1980s, during a period of economic crisis. Similarly, after the inauguration of the Madrid-Seville line the pace of investment in HSR slowed down. The more important point is that the massive concentrations of capital required by the development of the AVE and the problem of meeting them in periods of crisis is not a recent condition but rather a problem with an intricate history.

A commonly heard argument is that the development of HSR has been made possible by Spain's access to European funds. There is an important element of truth in this, as Spain has indeed been a privileged beneficiary of European development funds. But two simple facts will immediately alert us to the insufficiency of the observation. First, in practice, EU funds almost never exceed 25 percent of the total cost of any individual HSR project; second, the development of the AVE has not lost steam as Spain's access to EU funds grew more restricted (for details, see Audikana 2015). With this observation in mind it is easier then to turn our attention to that part of the funding structure which exists in the shadow of EU funds.

As visible in the structure of liberalization on the European level, the provision of railway services today embodies a dual relationship with regard to the question of monopolies. With regard to infrastructure provision, it is still widely believed that the best form to organize infrastructure provision is on a monopolistic basis. Service provision, however, should be reorganized in line with the objective of creating a single European market. However, EU policy impacts the actual development of infrastructure through several channels. An already mentioned one is the availability of EU funding. Importantly, though, on a national level, the question of HSR funding is tied to the broader question of the public deficit.

Throughout the post-Francoist history of the railways the question of the deficit of the railway companies has been ever present. This has also been essential to the way the problem of managerial autonomy of RENFE has been addressed. It has also been key to articulating the commercial orientation of railway services. The funding of HSR is likewise an essential part of this, as a key concern has been how to devise funding instruments that would not impact the national public debt. This has generated, broadly speaking, two types of solutions: (1) the extensive recourse to extra-budgetary funding and (2) the attempt to attract private capital through public-private partnerships. The recourse to extra-budgetary financing has involved the setting up of public agencies which would allow for debt financed development to appear on the balance sheets of companies without counting toward the national public debt. The establishment of GIF (*Gestor de Infraestructuras Ferroviarias*—Railway Infrastructure Manager) in 1997 was a crucial development in this respect. However, although this has been a provisional solution, it has constantly fired back as the problem of indebtedness of the companies themselves. To this should also be added the rising share of subnational level financing.

The policy of vertical unbundling that is the foundation of the liberalization model also has its origins in the problem of deficit management. In the 1990s when this became officially inscribed in the EU agenda, it was articulated as a response to the problem of the massive indebtedness of railway companies across Europe. As a matter of fact, this priority can be clearly seen in the fact that early EU policy required the separation of balance sheets, with the purpose of separating investment into infrastructure from service provision. In Spain the solution pursued was that of fully separating the companies, which is how RENFE Operadora and ADIF were formed, but other national companies resorted to maintaining the separation on the level of independent accounting. The objective of this separation was the already earlier formulated ambition of turning railway service provision into a commercially profitable activity. This is not a straightforward policy for the railways, however, since rail services are actually a bundle that includes goods that it is believed should be regulated and provided by the market, and those which are still considered as exceptions to the market-logic.

The space available in this chapter does not provide room for a broader discussion on the long history of this problem—namely, the malleable border between services that some believe should be entrusted to the market and those which are considered, for various reasons, outside of it. But one essential difference is that the separation between these services corresponds to the type of financing flexibility available on the governmental as well as on the company level. Those services that are believed necessary but cannot be reliably entrusted to the market are eligible for public subsidies. Such is the case of suburban rail and regional passenger transport, both of which are seen as providing an essential public service with social benefits. High-speed rail, initially exclusively a long-distance passenger service, was consequently not eligible for subsidies according to EU regulations. The underlying logic is that the operation of long-distance HSR services should be financed by the users. However, HSR in Spain has not managed to cover the cost of its operations through the model of the user-financed service. So, in addition to the massive scale of investments absorbed in the construction phase, the AVE has fallen short of its commercial objectives in the operational phase. With the estimated number of users lower for every new line opening, it is difficult to believe that this could change in any way in the foreseeable future.

It becomes clear, then, that the dispute between defenders and opponents of the AVE is not actually restricted to what critics would call the conflict between political planning and economic rationality. That is merely the ideological articulation of a vision of planning that aims to hide its own normative basis. Once we analyze this conflict, it is revealed that this has enlisted not only different promarket economic visions but also a range of rival techniques of calculation. Upholding the narrative of modernization through HSR requires the constant production of factual evidence that allows the integration of these contradictions into the teleology of progress.

Conclusion

This brief survey of the origins and development of the AVE and the main arguments that surround its defense and its criticism points us toward the challenges of converting an infrastructure into an asset (Birch and Muniesa, this volume). Spanish HSR shows the ideological and institutional prerequisites of converting a previously public service into a stream of revenue. However, the dominant direction in the analysis of infrastructure as an asset does not confirm the model of transition from state owned public goods to a private model of investment. The problem of infrastructure as an asset extends beyond the question of finance and provision and into the question of how an infrastructure enters the logic of market calculation.

What the debates surrounding Spanish HSR reveal is that it is not only the success of converting infrastructure into an asset that can provide an opportunity for extending the realm of market calculation. Both the critics and the defenders of AVE, in light of its perceived failures, converge around the belief that infrastructure should be either an outgrowth of market calculation or a foundational element in the construction of the market order. The problem of HSR as an asset can only be understood as a derivative of an effort to extend the realm of market calculation. Where the failure of turning HSR into a profitable asset becomes obvious, economic rationality becomes repoliticized in a manner that remains congruent with the project of marketization. To fully understand the challenges of turning HSR into an asset, it becomes necessary to surpass those views that would oppose the state and the market and derivatively treat private actors as privileged agents of assetization.

Notes

1. The completion of this chapter extends across my stay at two different institutions. The initial research was carried out as part of my research for a doctoral dissertation completed at the Central European University and was made possible by the support I received through a Wenner-Gren Dissertation Fieldwork Grant. The text was completed at the Max Planck Institute for Social Anthropology, where I have expanded my research on high-speed rail as a member of the *Financialisation* research group. Details have been modified in order to ensure anonymity of the interviewees.

2. Zapatero dice que los AVE refuerzan la cohesión territorial al inaugurar la línea Córdoba-Antequera, *El País*, 17 December 2006, http://elpais.com/diario/2006/12/17/andalucia/1166311327_850215.html.

3. Rajoy excluye a Zapatero de la inauguración del AVE a León, *El Español*, 29 September 2015, http://www.elespanol.com/espana/20150929/67743264_0.html.

4. The *Sindicato Español de Maquinistas y Ayudantes Ferroviarios* (SEMAF) is the Spanish Engine Drivers' Union and currently represents the majority of the train drivers.

5. SFF-CGT. 2001. Nuestro modelo de ferrocarril. Una alternativa de transporte social y seguro.

References

Albalate, D., and Bel, G. 2011. Cuando la economía no importa: Auge y esplendor de la Alta Velocidad en España. *Revista de Economía Aplicada* 55 (19): 171–190.

Albalate, D., and Bel, G. 2012. *The Economics and Politics of High-Speed Rail: Lessons from Experiences Abroad*. Lanham, MD: Lexington Books.

Audikana, A. 2015. L'Europe fait la grande vitesse, la grande vitesse fait l'Europe: l'influence européenne sur le développement de la grande vitesse espagnole. *Revue Géographique de l'Est* 3–4 (55), https://rge.revues.org/5533.

Bel, G. 2007. Política de transporte ¿Más recursos o mejor gestión? *Economistas* 111: 279–284.

Bel, G. 2010. La racionalización de las infraestructuras de transporte en España. *Cuadernos Económicos de ICE* 80: 211–228.

Bel, G. 2012. *Infrastructure and the Political Economy of Nation Building in Spain*. London: Sussex Academic Press.

Buier, N. 2016. "Time Is Not a Military Rank": Historical Memory and the Liberalization of the Spanish Railways. PhD diss., Central European University, Budapest.

Comín, F., Martín Aceña, P., Muñoz Rubio, M., and Vidal Olivares, J. 1998. *150 años de historia de los ferrocarriles españoles*. Madrid: Fundación de los Ferrocarriles Españoles y Grupo Anaya.

Davies, W. 2013. When Is a Market Not a Market? "Exemption," "Externality" and "Exception" in the Case of European State Aid Rules. *Theory, Culture & Society* 30 (2): 32–59.

Davies, W., and McGoey, L. 2012. Rationalities of Ignorance: On Financial Crisis and the Ambivalence of Neo-Liberal Epistemology. *Economy & Society* 41 (1): 64–83.

De Rus, G., and Nombela, G. 2007. Is the Investment in High Speed Rail Socially Profitable? *Journal of Transport Economics and Policy* 41 (1): 3–23.

De Rus, G., and Román, C. 2006. Análisis económico de la línea de alta velocidad Madrid-Barcelona. *Revista de Economía Aplicada* 14 (42): 35–79.

European Court of Auditors. 2018. *Special Report no 19: A European High-Speed Rail Network: Not a Reality but an Ineffective Patchwork*, https://www.eca.europa.eu/en/Pages/DocItem.aspx?did=46398.

Haughton, G., and McManus, P. 2012. Neoliberal Experiments with Urban Infrastructure: The Cross City Tunnel, Sydney. *International Journal of Urban and Regional Research* 36 (1): 90–105.

Hildyard, N. 2012. *More than Bricks and Mortar: Infrastructure as an Asset Class: Financing Development or Developing Finance?* Dorset: The Corner House, http://www.thecornerhouse.org.uk/resource/more-bricks-and-mortar.

Martín Baranda, G. 2011. *El AVE Madrid-Sevilla: crónica de una aventura*. Madrid: Ediciones Endymion.

Méndez, J,. De Rus, G., and Barrón, I. 2009. *El transporte ferroviario de alta velocidad: Una visión económica*. Bilbao: Fundación BBVA.

Muñoz Rubio, M. 1995. *RENFE (1941–1991): Medio siglo de ferrocarril público*. Madrid: Ediciones Luna.

O'Brien, P., and Pike, A. 2015. The Financialisation and Governance of Infrastructure. *IBUILD Working Paper No. 8*.

O'Neill, P. 2009. Infrastructure Investment and the Management of Risk. In *Managing Financial Risks: From Global to Local*, edited by G. Clark, A. Dixon, and A. Monk, 163–188. Oxford: Oxford University Press.

Porter, T. M. 1992. Objectivity as Standardization: The Rhetoric of Impersonality in Measurement, Statistics, and Cost-Benefit Analysis. *Annals of Scholarship* 9 (1–2): 19–59.

Porter, T. M. 1995. *Trust in Numbers. The Pursuit of Objectivity in Science and Public Life*. Princeton, NJ: Princeton University Press.

Ureña, J. M. 2012. *Territorial Implications of High-Speed Rail: A Spanish Perspective*. Farnham, UK: Ashgate.

6 Turning Sunlit Rooftops and Windy Sites into Energy Assets

Alain Nadaï and Béatrice Cointe

Introduction

In their introduction to this book, Birch and Muniesa define assets as things that "can be owned, traded, and capitalized as a revenue stream, often involving the valuation of discounted future earnings in the present." They insist that assets "are made" and are not "the consequence of some inherent or embodied quality." In charting dimensions of the asset form to be explored, they emphasize the legal, economic, and financial dimensions.

This chapter aims at advancing our understanding of the ways in and through which assets are made and what it means precisely to say that they are "made." In order to explore these questions, we start from two case studies of politically engaged renewable energy (ReN) development projects in France: a mutualized photovoltaic (PV) project—the Fermes de Figeac—carried out by farmers in the Lot department (Cointe 2016, 2018), and a wind farm "repowering" in a European migratory corridor involving bird watchers and a wind power developer in Narbonnaise, Languedoc-Roussillon department (Nadaï and Labussière 2010).

The current framing of renewable energy development by policies such as feed-in tariffs (FITs)—investment subsidies in the form of a guaranteed fixed tariff for a kWh of electricity generated from renewable sources—makes the financial dimension a central part of current ReN development. Policies such as FITs are designed to attract investment in the sector by guaranteeing a stable revenue stream in the mid-term future (e.g., twenty years or so).

Based on these two case studies, we empirically analyze the process of constituting qualities that allow parts of the environment—wind, sun, a roof, a site—to become owned, capitalized as a revenue stream, and possibly traded. In particular, we explore the territorial, spatial, and political dimensions of this making and their articulations in forms of calculation, aspects that have not been much covered by the literature interested in assets and assetization processes (see Buier, this volume).

At the core of ReN development is the work of turning things such as sunlit roofs or windy sites into productive entities—solar roofs or wind farms producing ReN kilowatt hours—through the siting of material devices (PV panels, wind turbines) (Nadaï et al. 2018). The resulting entities convert an untamed flow (sun, wind) into electricity dubbed "renewable" and, in the current French policy context, eligible for stable remuneration (feed-in tariff). They produce value in the form of marketable electricity. They are thus valued in different ways. For instance, investors may regard solar roofs or wind farms as potentially profitable opportunities to invest in. In certain cases, they may gauge the financial viability of a ReN developer by considering its ongoing (productive) solar/wind farms portfolio. Sometimes solar or wind farms can even be traded over the counter by ReN developers in order to fine-tune their project portfolio and improve their worth for potential investors. Like assets, they thus articulate different types of revenues—that is, profits and rents (Birch and Tyfield 2013).

The work of assembling productive-enough ReN entities commonly falls under the heading of project development. The project has a multiple existence as a social, technical, territorial, financial, legal, and regulatory entity. A successful ReN project incorporates and articulates all these dimensions together. For example, the siting of wind turbines must be rendered compatible with ongoing uses of land, landscape, and air and with administrative requirements. The scaling of the project must allow for sufficient return on investment in order to attract investors and to permit the financing of the project. In certain cases, project design must allow for sharing the value derived from ReN production.

Hence, developing a project requires engaging things—roofs, sites, turbines, PV panels, but also wildlife, landscape, and so on—in hybrid "agencements" by requalifying their preexisting usages and/or users. Callon (2008) defines an agencement as a hybrid ensemble; it foregrounds both materiality and the fact that agency is distributed and derives from the relations among hybrid entities. In what follows, we do not approach assetization from the exclusive perspective of financial calculation, which is only a part of the work of financialization (Chiapello 2017). Drawing on Leyshon and

Thrift (2007), we are interested in the articulation between the agencing of productive activities and the work of financialization (see also Gilbert, this volume). We thus look at ReN projects development and consider the extent to which the requalifications at work in these developments bear the imprint of financialization and capitalization, conceiving the latter as a multidimensional, cultural process (Muniesa et al. 2017).

Our case studies allow us to describe the socio-technical and economic reconfigurations through which sunlit rooftops and a windy site are turned into ReN projects and made to generate revenue streams for the developers of these projects and/or for the owners of the roofs/plots of lands. This allows us to discuss three aspects of assetization and the asset form: (1) the relational work through which assetization articulates different entities and types of values, financial, territorial, and political, leading to the suggestion of talking of *assets-as-agencements*; (2) the specific role of the state (through feed-in tariffs) in setting the conditions for entities such as wind turbines /sun roofs and sites to be engaged in assetization processes; and (3) the uneven extent to which the work of financialization underpins that of assetization.

The chapter is structured as follows: the first part sets the analytical frame, the second and third parts present our case studies, and the fourth part draws on these presentations in order to discuss the three insights this chapter brings to the discussion about assets forms—assets as agencements, the role of the state in triggering assetization processes, and the articulation between financialization and assetization.

Assets as State-Triggered Agencements

As developed in the introduction to this book, several dimensions of assets have been outlined in the literature. Within the wider literature, three sources of inspiration are important for this chapter.

The first is the work of Leyshon and Thrift (2007) proposing to address the way in which financialization may steer and reframe mundane economic practices and also the spatial, global, geographical dimension of assetization. The authors study the search for new, sometimes unexpected, spaces that can yield dividends and be constructed as assets, a proposal which resonates with the search for windy or sunny areas in our case studies. The authors emphasize securitization—the pooling of ordinary sources of income along dimensions of risk and rewards—as the way capitalists have to construct predictable income streams and construct ordinary activities as collateral in their search for capital. Entities that were there as mere sources of revenues, not worth capitalization, become assets through their requalification (assessment/sorting), pooling, and articulation within the financial system. Financialization can end up imposing conditions and requirements on the way in which these sources of economic revenues are managed.

The work of Leyshon and Thrift is inspiring because of its emphasis on the geographical dimension of assetization (the search for new spaces for development). Yet our analysis does not focus on the global geography of financialization. We target instead a local, spatial dimension of assetization, which has not been addressed in the literature. Similarly, rather than securitization, our case studies involve only some type of pooling. The construction of predictable income streams mainly operates through feed-in-tariffs, which trigger investment in the development of concrete economic activities (ReN projects) with simultaneous productive, territorial, and political dimensions. Hence, central to our exploration are the relations between these different dimensions and the multiple values associated with them, and the relations between these values, the financial dimension, and the role of state public policy (see Milyaeva and Neyland, this volume; Williams, this volume).

Chiapello (2015, 2017) has described the work of financialization, including that of public policy. She shows how the financialization of public policy may change the definition of domains of activities, objects and professional practices: what she calls "colonization"—that is, when the values and practices of financialization penetrate deep into the core of practices, values and meaning. In so doing, she points to the upstream expert work of explicitation (Muniesa 2014; Muniesa and Linhardt 2011), which interprets these domains of activities, objects, or professional practices in terms of risks to be quantified and monetized so as to translate them in terms of investment choices and profitability. Here again, Chiapello mainly insists on the relation between financialization and the language and categories of accounting, while our case studies foreground territorial and environmental dimensions. They thus invite us to explore the way in which explicitation and/or colonization amounts to a work of hybridization between heterogeneous values and entities. This connects our analysis with an argument in STS about marketization processes involving multiple values and valuations (e.g., Callon 2008; Callon et al. 2013; Geiger et al. 2015), and shows how assetization may gain in being approached as a relational work of agencement.

Last but not least, as feed-in-tariffs foreground the central role of the state in channeling money flows toward certain things (renewable energy production), our case studies offer an occasion to reflect on the role of the state in setting up the actual flows of money necessary to make financialization happen. Boltanski's and Esquerre's (2016) work about the "economy of enrichment"—an economy of constructing value from and around already existing objects—offers a good perspective for reflecting on these matters. The authors characterize assetization as a process of constructing new value out of already existing objects by engaging them in a future-oriented narrative. The economy of enrichment then plays with difference (between these objects and others) and time (horizons, future revenues) to enrich these objects. Interestingly, analyzing the setting up of feed-in tariffs along these lines allows us to specify the role of the state in differentiating ReN from other resources and in channeling money flows to ReN projects to make their future value predictable. This has two advantages. It allows us to link the limited need for financial calculation in our case studies (e.g., no genuine securitization) to the state-backed promise of value. It is also an invitation to understand better the politics allowing for such a promise of value and for its implementation in actual money flows.

Mutualizing Sunshine

Our first case study is a photovoltaic (PV) project initiated in 2008 by an agricultural cooperative in southwest France, the Fermes de Figeac. It was made possible by feed-in tariffs for PV-generated electricity, which were very high in France in 2008 and thus promised high rates of return. At the same time, the project took part in the cooperative's broader strategy of territorial innovation and development. Feed-in tariffs were particularly high for building-integrated photovoltaics: large sunlit rooftops thus became a potentially profitable resource. The cooperative's idea was to pool rooftops owned by farmers in the area and to install photovoltaic systems on them to constitute a scattered photovoltaic park that would be managed in a mutualized way. The objective was to yield income—for farmers, for the cooperative, and for reinvestment in other territorial projects—from a resource that feed-in tariffs suddenly made financially interesting, but in a way that reassured the cooperative in its commitment to mutualization and helped revive the territory. As long as the resource was there and gained the attention of various project developers, the cooperative (and many farmers) thought it better to exploit it in a way that kept profits local.

A company was created specifically for the project. Its capital was constituted by shares held by participants in the project (each participant brought in 20 percent of the amount needed to equip their rooftops with photovoltaics) and by funds borrowed in the form of a syndicated loan. The company signed leases to rent the rooftops and paid for the installation of photovoltaic systems. It also took care of all the administrative and technical procedures (building permits, purchase agreements for the sale of electricity at the feed-in tariff, grid connection). The electricity generated is injected into the grid and sold to EDF (Électricité de France, the dominant French electricity utility company) at the feed-in rate that was in place in 2008 (60 €c/kWh). Income is redistributed to shareholders via rents paid for their rooftops and dividends, with each square meter of photovoltaic panel considered as yielding the same amount. On the balance sheet, roof surfaces were converted into shares (i.e., in investments).

A computation of the average solar radiation in the area and of the feed-in tariff rate projected provisional income, expecting profits after eight years in operation (so from 2017 on). The company is set to be discontinued after twenty years, at the end of the purchase agreements, and the revenue stream was projected until this date. Once the park was installed, the cooperative organized a system for maintenance, hiring an engineer and a technician, and devising software and a communication system to monitor the functioning of the park as a whole. This system of proximity maintenance is a way to maximize profits by ensuring that the park works as well as possible all the time. With this project, the cooperative combined a logic of borrowing and investment—projecting income, assessing risks, aggregating rooftops, negotiating with bankers (who had their own procedure for risk assessment)—with a territorial approach to turn rooftops into a source of monetary value and a source of new dynamism for the territory. The commitment to local development was explicit in the design of the project, since the company's status prevents shares being transferred to "outsiders" (i.e., to anyone without direct links to the rooftops that are part of the project).

Sharing the Wind

Our second case study is a wind power repowering project located in La Narbonnaise Parc Naturel Régional (PNR: "regional nature park"), at the border between France and Spain on the east side of the Pyrenean mountains. As one of the windiest places in France, Narbonnaise attracted early interest for wind power development. It was the place where the first, very early industrial wind power project was built in France in 1990, with EU Thermie funding (before any feed-in tariffs were in place). In 2010, a new project proposed the first repowering project in the country (i.e., dismantling the wind farm in order to set up a new one). Neighboring communities shared an interest in repowering the wind farm located on common land, which allowed them a part of the revenues from the farm (taxes, land rents).

As a windy place, the small littoral plain of Narbonnaise is also one of two migratory routes for birds on their way from Africa to Eastern Europe and back. Narbonnaise has a strong political history of bird watching. It was one of the most important places in France where bird watchers met and set up "migration camps" in the 1970s, to attract Europe's and the French state's attention to the traditional hunting of birds of prey, endangered species, and the need for regulatory protection. Migration camps notably politicized birds by counting the population of birds passing through.

The wind power site of this second case study is thus located within a major migration corridor, which made birds an important project adjuster. The project has a somewhat standard financial approach, through the setting up of a private project company by the developer, but is original in two ways. First, its design and siting proposal involved a collaboration between the wind power developer and the local branch of the French bird protection organization (Ligue pour la Protection des Oiseaux, or LPO). Second, part of its benefits were to be shared with local actors—the LPO, a local hunters NGO, a local shepherd—in order to allow for environmental compensation and for the monitoring of the impact of the future wind farm on migrating birds.

As a collaboration between developer and bird watchers, the project design triggered an experiment. The LPO developed a new method for observing birds and connected it with its national strategy in the area of wind power planning and its European network of bird protection NGOs. "Micro-siting," as this method is called, focuses on the way in which birds use a site, including the wind and the (eventual) presence of wind turbines. Unlike usual bird watching methods in the Narbonnaise area, micro-siting is not a census nor is it primarily about counting the size of the migrating species populations. It is about following individual birds so as to understand the way in which they develop strategies in relation to the presence of wind turbines. It is about individual stories, individual or small group successes, difficulties or failures in passing through, beside, or over a wind farm. It is about birds' strategies in finding thermals and updrafts, about soaring and gliding. By focusing on individual stories so as to capture birds' intelligence, micro-siting multiplies observations and expands statistical reach, which allows it to translate birds' strategies into indices that are congruent with planning and siting practices. It thus endows birds with new capacities and ventures in evolving the politics of bird protection in the same area where migration camps had politicized it in the 1970s.

With this project and experiment, the developer and the LPO opened access to spaces that were deemed protected because of bird migratory movements in order to turn them into energy productive spaces.

Assetization and the Asset Form

Both case studies foreground the work of assembling productive agencements, which is at the core of project development, and the values that are associated with this work. This allows us to make three main contributions to the discussion about asset forms and the making of assets.

Assets-as-Agencements: A Relational Work

Both case studies show that multiple values are associated with project development. Hence assetization is not restricted to the economic, financial, or legal domains, which is the main focus of the literature (see Birch and Muniesa, this volume). On the one hand, the feed-in tariff model can be regarded as a penetration of market and finance in the sense of a stable remuneration (tariff for twenty years) attached to a stable electricity commodity (a standard kWh injected into the grid, a tradable certificate of "renewable origin") with the objective to trigger investment in and capitalization on ReN projects. Feed-in tariffs offer an opportunity for profit, provided that sun or wind can be turned into standard electrical kWh thanks to electricity producing devices (PV panel, wind turbine). Roofs and sites that were not regarded as value-generating places suddenly become potential sources of income. On the other hand, value making is politically driven. First, because of the various administrative procedures, which make a certain environment or landscape (environmental assessment, visual assessment, physical resource assessment) matter in the process of project development. Second, because of values—such as mutualization in the case of Figeac, environment or bird protection in the case of Narbonnaise—which are shared by the actors engaged in these specific processes. Both processes go beyond what is usual or usually required by administrative procedures for the development of renewable energy projects. They can thus be said to be political in the sense that they overcome usual policy frameworks, point to specific issues and invent their way of dealing with project development (Barry 1999).

Nonetheless, in both case studies, the attention and work devoted to project development in the collective or environmental dimension make explicit the *ontological dimension* of this development. What we mean here by ontological stems from a relational approach that sees entities as the product of relations and practical activities (Simondon 1989; Woolgar and Lezaun 2013) and enables research to shed light on ways through which things can be requalified in order to seem the same (Mol 2002; Law and Lien 2013), and to be invested with political or normative capacities (Marres 2012). The productive entities that emerge through these processes do not just connect preexisting entities. They amount to a new relational realm, weaving together what is there (roof, sites), with new value making and sharing devices which include the feed-in tariff, know-how (bird watching techniques, bird classification, affects, mapping, aeolic grid, knowledge of local variations in sunshine) and material devices (PV panels, meters, ICTs, PV panel-cleaning robot, on-site wind turbines, bird watching devices, etc.). By bringing what is there into a new net of relations and qualifications, the new agencement allows entities to become active in a different way, as energy-producing entities.

In Figeac, a first step consists in actively recruiting farmers potentially interested in taking part in a PV project and in assembling a pool of roofs and a collective that can be equipped to perform as a single mutualized project. Material devices such as meters and ICT connection with the PV-farm technicians allow a dedicated team within the cooperative to follow production in real time and to engage farmers in the collective monitoring and performance of the photovoltaic park (they are asked to intervene or check in case of problems). Farmers thus become energy producers. This organization is a crucial part of guaranteeing the stability of the income generated by the rooftops. It also translates into balance sheets: actual production is recorded for each installation and compared against theoretical production. Anomalies in electricity production for individual roofs are made visible as differences from the prevision, so that it takes only a few steps to translate them into financial gain or loss (compared with expectations). Roof leasing contracts and share-holding in the PV project (a simplified joint-stock company) turns not only previously unproductive private roofs into shared sunlit productive units but also turns farmers into investors, shareholders, and renters. Share-ownership, however, is bound to the buildings equipped with photovoltaics: only people related to these buildings are entitled to hold shares, and shares cannot be sold without the PV company administrative council's agreement.

In Narbonnaise, the major challenge for gaining a renewed access to the wind is to gain access to the site as a migratory corridor. Reactivating the site as an energy productive entity requires an exploration of the potential for compatibilities between migrating birds and new wind turbines on a finer spatial grain, a novelty at that time. Existing wind turbines are turned into a type of lab-scape for the experiment. Bird watchers hide behind the turbines, observe and follow individual birds in their crossing through the existing wind farm, gauging their individual cognitive/strategic ability to pass through, over, under, or beside the working turbines. The reshuffling of bird classifications (according to crossing ability rather than to statutory protection) and the drawing of ensuing individual bird trajectories both allow for the mapping of migratory micro-corridors, which paves the way for wind farm siting proposals judged compatible with bird migration.

While this opens a new potential for sharing the wind, it raises accountability issues on the part of bird watchers (it requires a follow-up) for birds not to be unduly put at risk. On the part of the wind power developer, this implies a limitation in the size and power of the new wind turbines in order to allow for birds migrating conditions. In this process, entities are not only requalified but also perform differently. Birds overall become more skilled (wind farm compatible). The wind farm overall becomes more compatible with migrating birds. The siting of the turbines compromises between profit maximization and the safety of migrating birds. Its revenue is also partly recycled into onsite bird watching in order to endorse accountability, and partly directed in support of herd grazing in order to improve the habitat for small game, which are prey for raptors and for local hunters whose associations are engaged in a joint EU Life Project. Changes are therefore of two kinds. Some of them are scripted in the very materiality of the project and cannot be changed were the project to be sold to another developer in the future. Others are contractual agreements that can potentially be revised in such a prospect.

In both case studies, processes lead to hybrid collectives (farmers/cooperative/energy producers; bird watchers/hunters/wind power developers), hybrid roofs (private homes/energy production homes/revenue earning devices) and hybrid sites (wildlife habitat/agricultural commons/energy production). *Hybridization here is thus a way of accessing spaces* in order to access resources (sun, wind). Spaces that were not producing energy are turned into energy productive spatial agencements (PV roofs, wind farm). Our analysis thus contains a genuinely *spatial dimension* that we have not seen in other analyses of assetization.

Here "accessing" has a double meaning. It means colonizing spaces in the sense of expanding into or occupying spaces, but also colonizing in the sense of changing ways of perceiving, representing, practicing, and performing (Chiapello 2015). These are inseparable, since ontological requalification allows for hybridization, which in turn allows for the spaces to become shared and productive spaces. Roofs (for the collective of farmers in Figeac) and sites (for the neighboring communities in La Narbonnaise) become sources of a stream of future revenue because the set of relations in which they are engaged is changed. The relation of the farmers to their roofs is changed (from property to roof property + PV-roof rental contract + part-ownership of the PV renting entity). The relation of the birds to the site is changed (from protected species/protected site to skilled species/spatially differentiated micro-corridors).

While such requalifications open a spatial access to the resource (sun or wind), they do not imply access to the (future) value set by the feed-in tariffs. The articulation between access to the resource and access to the value set by the tariffs in terms of monetary flows is taken charge of by an organizational and financial agencement: the project company. Neither the roofs nor the site are capitalized as such; neither can be traded as energy producing entities, neither receives flows of money coming from the French state. What receives money in exchange for kWh injected into the grid, what is capitalized, what is potentially traded (but *not actually traded in our case studies*) are the projects—that is, the project companies: the legal entities associated with this materiality and spatial access.

A project company can endorse a variety of forms and status that we cannot detail here (see Poize 2015; Médiation & Environnement 2016;

Poize and Rüdinger 2014). The simplified joint-stock company adopted in Figeac is only one among many possibilities. Forms and status allow for very different articulations between the collective of actors financially participating into the project, as money lender, shareholder, beneficiaries, and the material, spatial, territorial anchorage of the project. Financial engagement can be tightly tied to material, spatial, territorial participation, as in Figeac, where only roof holders are entitled to hold shares and shares cannot be sold without the PV company administrative council's agreement. In other renewable energy projects, however, financial engagement is kept completely separate from the other dimensions of the project, the developer being the sole project manager.

While project companies can endorse many statuses, they always allow for various types of revenues. One is the discounted income flow as payment for upcoming electrical production. This income can be redistributed among the (varying) parties in the form of environmental compensations (financial support to herd grazing, birds follow-up in the case of Narbonnaise), rents (rooftops, communal lands), per production payment (to the developer or mutualized as in Figeac) through fixed tariffs covering development and exploitation costs, and the remuneration of developers (IRR) and shareholders (IRR shareholders). As emphasized by Birch (2017) and Muniesa (2012), capitalization and assetization are undertaken through such organizational settings and management practices.

Hence assetization is the outcome of relations that can be arranged in different manners. In our case studies, things that become productive are engaged into *agencements-as-assets*. In the examples developed by Leyshon and Thrift (2007), it is the creation of secondary markets and the securitization of assets that turn the original market (real economy) into an asset that can be used as collateral for financial operations (capitalization). In our case studies, roofs or sites are not productive in themselves: they become productive by being woven into a network that connects them to a project company, the electricity grid, the feed-in tariff, and so on. It is thus the project company that makes the site or the roof active and worth capitalization. The extent to which financialization impacts the way in which the roof or the site is rendered productive depends on the type of interweaving that the project company settles upon.

Considering assetization as the outcome of relations that can be arranged in different manners through agencement thus seems important if we want to make sense of the significant differences in the relations between projects and finance.

Enriching, the Decisive Role of the State

In many respects, FITs can be considered as capitalization devices in that they are devices which potentially turn a "variety of things… into [assets] that have the power to generate streams of future revenues" (Doganova and Muniesa 2015). Yet they do so in a peculiar way compared with all the examples of sophisticated valuation processes that have been described, such as business models and IPRs, or discounted cash flows calculation in genetic engineering, or pharmaceutical R&D (Doganova 2012), or securitization (Leyshon and Thrift 2007).

With FITs, defined as political and economic arrangements (Cointe, 2014), governments decide and set the value of the future flow of revenue per ReN kWh sold, leaving uncertainties only about the scaling of individual ReN projects and whether or not these will be able to access the spaces (and wind or sun resource) that will make them profitable. Hence, for entrepreneurs, FITs simplify the valuation process to a great extent. Further, with renewable energy sources, especially unavoidable energy sources such as sunshine or wind, the amount of electricity generated can be projected with relatively low uncertainty. For instance, the business plan of the Fermes de Figeac's project relied on two simulations taking into account only two sets of assumptions: average sun exposure and interest rates. The need to undertake complex calculations, to perform a scale or role model, is greatly simplified because the main issue becomes demonstrating that the project will be able to access the space that makes it profitable—that is, both convincing the necessary actors on the local level and gaining the administrative and technical authorizations required. Differently stated, because the value is constructed politically, calculations translating entities in terms of risks and rewards to demonstrate the stability and value of a future income flow are not what requires the most attention. What is crucial is the construction and demonstration of access, chiefly through the very practical arrangement of material and administrative entities. As we have indicated, this work is akin to that of explicitation which implies ontological requalifications while not directly implementing them along (explicit) financial calculations.

On a broader level, FIT can be regarded as underpinning assetization by enriching renewable energies through a play with difference and time

(Boltanski and Esquerre 2016). The very existence of FITs bears witness to a state's work of (1) setting a difference between renewable energies and other energies, (2) setting the value of renewable energies and stabilizing it over time, and (3) channeling actual money flows to ReN kWh as a delimited object (commodity). As Boltanski and Esquerre point out, making money converge in an object (ReN kWh) makes this object (ReN kWh) generate money (money-commodity-money), turning it into an asset.

Existing analyses of the career of feed-in tariff in France and the European Union (Cointe 2014, 2015; Cointe and Nadaï 2018) testify to the upstream political work that has been required of member states in order to set the tariff in motion. Member states had to legitimate feed-in tariffs as a renewable energy policy instrument before the European Commission, which was desiring to develop the EU as a market economy. This first supposed a differentiation of fossil and nonfossil energies with respect to their environmental and social benefits. It also required making the tariff congruent with the integrated electricity market vision by playing with both tariff design (making it more market-like—that is, responsive to demand and calculation based on externality assessments) and the very definition of what ought to be considered as the (market) value of renewable energies (playing with ideas of costs, such as avoided cost of production, avoided externality, production cost). This process attests to the mutual entanglement of the political and economic orders behind the attribution of a predictable value to ReN production. In addition, member states had to legitimate the necessary money transfer underpinning the implementation of any feed-in tariff on a national level, ranging from electricity consumers, tax payers, or fossil energy producers to ReN developers. In France, this transfer was implemented by raising a tax already borne by electricity consumers. Finally, the tariff was implemented through state enforcement of a mix of property and contract law. While the resource (sun, wind) remained *res communis*, the ownership of its energy ended up being governed by that of the technical device harnessing it (PV panel, wind turbine). Only the owner of this device can contract for the tariff, provided they succeed in getting a construction permit (access to the resource) (Nadaï and Labussière 2017; Nadaï et al. 2018).

As a consequence, the construction of a predictable income associated with renewable energy results from a construction that operates on multiple levels. The very existence of a tariff results from meta-calculation and political and institutional negotiation. It legitimizes a founding difference between renewable energies and other energies, and the possibility of stabilizing a value for the former in the future. These are preconditions for assetization to emerge, as suggested by Boltanski and Esquerre (2016). Still, they are not sufficient. The evolving value of a given feed-in tariff (always contracted at a-point-in-time-value for an entire period—e.g., 20 years) mirrors a state's specific objectives for the development of new energies. Important variations may be apt to unsettle the predictability of ReN value. At the beginning of the 2000s (Haas et al. 2004) or more recently for the French PV tariff (Cointe 2015), the instability of renewable energy policies was pointed out as one reason for their slow development. Hence, the impression given by our case studies that there is no need for such complex proof of value or calculation as there is in classical examples of financing innovation (Doganova 2015) is a result of the calculations being the clear product of a multilayered political construction. This construction sometimes succeeds in establishing stable enough time-difference configurations to enrich renewable energies and allow for the relational process leading to their assetization.

Assetization and Financialization

With respect to the financial dimension, our case studies are also specific in two ways. First, because of FITs, there is no (or little) uncertainty as to the future revenue stream that can be derived from these projects once their initial development phase has begun and they have received the various administrative authorizations. So again, as opposed to studies of innovation-based start-ups (Doganova 2012, 2015), what entrepreneurs need to prove to investors is not that there is potential future value, but that they are able to access it—to get the construction permit and grid-connection authorization, to fund the installation of ReN production devices, and to site their project materially. The fact that the future value is defined by public policy to a certain extent alleviates the pressure of financial considerations on project development. Investors can still attempt to play with cost reduction or the scaling of the project in order to increase the project rate of return on investment (RRI), but there is no room or need for a race to the bottom on unit kWh remuneration, which limits the pressure on development costs.

Second, these case studies can be regarded as innovative ReN projects for France (Nadaï et al. 2014) in that they are based on the participation of politically engaged actors—the Figeac agricultural cooperative (mutualization, territorialization), the LPO (bird protection, environmental protection). In

France, most wind power or PV solar projects are developed by private ReN developers. While developers need to take into account local configurations in order to get access to sun (roofs) or wind (site), profit-making and financial values tend to be more prevalent in project development processes than described here, with less attention and work devoted to the collective or environmental dimension of the projects (see Debourdeau 2011).

Hence, in both cases financialization is limited as a consequence (intended or unintended) of the design of the project. Entities—birds, roofs—are assessed, sorted, and aggregated, but only to a certain extent. While bird protection is reprocessed as to its risks, the gains that ensue from protecting (or not protecting) bird lives is not translated into financial terms nor even into monetary terms. Of course, there is a gain expected by the developer in collaborating in the processing of birds in relation to their risk—the gain in accessing the tariff by getting a construction permit for the wind farm. Yet the gains in protecting birds are not processed through financial calculation in terms of their amount: bird protection is not reprocessed as to its rewards.[1]

In Figeac, there is a pooling of roofs according both to the technical and geographical suitability of roofs as sites for photovoltaic electricity generation and to the contribution of individual farmers to the capital of the project. Holding shares and investing in the project was a requirement for participation, but while there was financial consideration of these investments and their expected yield, the pooling of the farmers' collective and the roofs did not occur according to a computation of financial risk and reward (but rather according to the viability of the project), and the project sought to work with a relatively homogeneous set of roofs in terms of PV potential, contribution to capital, administrative requirements, grid connection cost, and feed-in rate. In a later step, the bank undertook the usual due diligence, disaggregating (roof by roof) and reaggregating (for the whole project) risk assessment, but this had no consequences as to the perimeter of the project. In fact, the main financial risk was that access to the resource (both feed-in tariffs and sunshine) would not be secured, so that little uncertainty remained once the installation was completed.

In neither case, therefore, was pooling run along a genuine reward dimension as described by Leyshon and Thrift (2007) or Eve Chiapello (2017) in their descriptions of the work of financialization. In these cases, we do not see any genuine financial calculation carried out in order to requalify or aggregate entities. Further, while the projects involve financial calculations—in Figeac, profits depend on sunshine, and electricity production is recorded as part of the accounting—this does not colonize all the material practices considered: bird life is not monetized, and access to shares is not solely based on financial contribution but rather grounded in local land ownership.

Assetization of renewable energy sources takes place through the setting up of project-specific companies that assemble the roofs or the site and carry out the work of turning them into productive entities. This, and the fact that the income stream yielded by projects is guaranteed by public policies (FITs), echoes Leyshon's and Thrift's examples of the Private Finance Initiatives (Leyshon and Thrift 2007, 105–106). In our cases, however, no secondary markets are constituted on the basis of the stream of income generated by the projects. Assetization is not carried all the way through, since assets are not traded, bundled-up, and incorporated in the international financial system. This owes to the specificities of our case studies but not to the specificities of renewable energy projects per se. It thus raises questions about the extent to which renewable energy projects are actually traded, incorporated in portfolios, or used to back further financial activities or speculation. This would call for a complementary analysis of renewable energy projects carried out by more "standard" developers. Moreover, it goes to show that the asset form is not an absolute and inherent quality (see Braun, this volume). It is rather a state that depends on the net of relations through which assetization is achieved. It can thus be restricted to a specific spatial and political site, as in our two cases, or carried forward by making assets transportable and widely tradable, depending on how assetization is performed.

Conclusion

This chapter has explored the territorial/spatial dimension of assetization processes, a point that has not been covered in the literature about assets. Our analysis shows that assetization cannot be reduced to a process of turning a well-delimited object or entity into a financial product. Instead, it is a process that associates a future value and revenue stream with an object or entity by including it in an agencement. Such relational work has an ontological dimension: it recomposes heterogeneous entities and values together so that they advance and contribute to a shared end. To that extent, things are engaged in assets-as-agencements.

The chapter describes case studies in which the state has a decisive role in setting a future value and in channeling money flows toward assetized entities. Building on Boltanski's and Esquerre's (2016) analytical proposal for an economy of enrichment, and based on the example of the feed-in-tariff for ReN in France and the EU, the discussion has emphasized the political work required from the state to stabilize a prospect for value and a predictable revenue stream, and attach it to given entities. We emphasized the specific tension raised in doing this within a purported market economy of which assets and assetization are supposed to be part. The tension runs between the work of differentiation allowing for money flows to be targeted at certain things (and not others) and the funding pledge for undifferentiation that underlies the political ideal of free and open competition.

This construction sometimes succeeds in establishing a stable enough time-difference configuration to enrich certain entities and allow for the relational process leading to their assetization to occur. We have shown that this construction then operates on multiple levels, suggesting that assetization processes are multilayered processes. The founding time-difference configuration associated with the adoption of a feed-in-tariff in a country is only a precondition for assetization: it provides only a breeding ground but no guarantee that the relational process engaging things in assets-as-agencement will actually take place.

When this is so, our case studies have shown that the need to prove the future profitability of assets and the work of financialization might be significantly alleviated. While financialization is still at work (for instance, through inscription in the balance book, since the founding difference and the future value have been set at a political/policy level), actors have to prove only that they can access the resource, which significantly reduces the work of calculation. This suggests that asset forms and their construction might be very diverse, and that more analyses of asset forming processes are needed.

Acknowledgments

This work was supported by the French Agence Nationale de la Recherche (French National Research Agency [ANR]) under the program "Sociétés innovantes" (grant number 2011-SOIN-003–01, project "COLLENER").

Note

1. This would anyway make little sense as the value of the tariff itself results from a meta-calculation that does not really take the externality of renewable energies into account. The idea behind its value is rather to allow investment to happen, and so to match a sufficient return on investment.

References

Barry, A. 1999. Demonstrations: Sites and Sights of Direct Action. *Economy and Society* 28 (1): 75–94.

Birch, K. 2017. Rethinking Value in the Bio-economy: Finance, Assetization, and the Management of Value. *Science, Technology, & Human Values* 42 (3): 460–490.

Birch, K., and Tyfield, D. 2013. Theorizing the Bioeconomy: Biovalue, Biocapital, Bioeconomics or … What? *Science, Technology, & Human Values* 38 (3): 299–327.

Boltanski, L., and Esquerre, A. 2016. The Economic Life of Things. *New Left Review* 98: 31–54.

Callon, M. 2008. Economic Markets and the Rise of Interactive Agencements: From Prosthetic Agencies to Habilitated Agencies. In *Living in a Material World*, edited by T. Pinch and R. Swedberg, 29–55. Cambridge, MA: MIT Press.

Callon, M., Akrich, M., Dubuisson-Quellier, S., Grandclément, C., Latour, B., Mallard, A., Méadel, C., Muniesa, F., and Rabeharisoa, V. 2013. *Sociologie des Agencements Marchands: Textes Choisis*. Paris: Presses des Mines.

Chiapello, E. 2015. Financialization of Valuation. *Human Studies* 38: 13–35.

Chiapello, E. 2017. La financiarisation des politiques publiques. *Mondes en Développement* 178: 23–40.

Cointe, B. 2014. The Emergence of Photovoltaics in France in the Light of Feed-In Tariffs: Exploring the Markets and Politics of a Modular Technology. PhD diss., EHESS, Paris.

Cointe, B. 2015. From a Promise to a Problem: The Political Economy of Solar Photovoltaics in France. *Energy Research & Social Science* 8: 151–161.

Cointe, B. 2016. Le tarif d'achat photovoltaïque comme outil d'innovation territoriale: l'exemple des Fermes de Figeac. *VertigO* 16 (1) (May), http://vertigo.revues.org/17040.

Cointe, B. 2018. Mutualising Sunshine: Economic and Territorial Entanglements in a Local Photovoltaic Project. *Local Environment: The International Journal of Justice and Sustainability*, doi:10.1080/13549839.2018.1436044.

Cointe, B., and Nadaï, A. 2018. *Feed-in Tariffs in the European Union: Renewable Energy Policy, the Internal Electricity Market and Economic Expertise.* New York: Palgrave Macmillan.

Debourdeau, A. 2011. De la "solution" au "problème": La problématisation de l'obligation d'achat de l'énergie solaire photovoltaïque en France et en Allemagne. *Politix* 24 (95): 85–109.

Doganova, L. 2012. *Valoriser la science: les partenariats des start-up technologiques.* Paris: Presses des Mines.

Doganova, L. 2015. Que vaut une molécule? Formulation de la valeur dans les projets de développement de nouveaux médicaments. *Revue d'anthropologie des connaissances* 9 (1): 17–38.

Doganova, L., and Muniesa, F. 2015. Capitalization Devices: Business Models and the Renewal of Markets. In *Making Things Valuable,* edited by M. Kornberger, L. Justesen, J. Mouritsen, and A. Koed Madsen, 109–215. Oxford, Oxford University Press.

Geiger, S., Harrison, D., Kjellberg, H., and Mallard, A., eds. 2015. *Concerned Markets: Economic Ordering for Multiple Values.* Cheltenham, UK: Edward Elgar.

Haas, R., Wolfgang, E., Huber, C., Langniss, O., Lorenzoni, A., Madlener, R., Ménanteau, P., Morthorst, P. E., Martins, A., Oniszk, A., Schleich, J., Smith, A., Vassk, Z., and Verbruggen, A. 2004. How to Promote Renewable Energy Systems Successfully and Effectively. *Energy Policy* 32: 833–839.

Law, J., and Lien, M. E. 2013. Slippery: Field Notes on Empirical Ontology. *Social Studies of Science* 43 (3): 363–378.

Leyshon, A., and Thrift, N. 2007. The Capitalization of Almost Everything: The Future of Finance and Capitalism. *Theory, Culture & Society* 24 (7–8): 97–115.

Marres, N. 2012. *Material Participation: Technology, the Environment and Everyday Publics.* Basingstoke, UK: Palgrave.

Médiation & Environnement. 2016. *Quelle intégration territoriale des Energies Renouvelables Participatives?* ADEME Research Report. Paris: ADEME.

Mol, A. 2002. *The Body Multiple: Ontology in Medical Practice.* Durham, NC: Duke University Press.

Muniesa, F. 2012. A Flank Movement in the Understanding of Valuation. *Sociological Review* 59 (2): 24–38.

Muniesa, F. 2014. *The Provoked Economy: Economic Reality and the Performative Turn.* London: Routledge.

Muniesa, F., Doganova, L., Ortiz, H., Pina-Stranger, A., Paterson, F., Bourgoin, A., Ehrenstein, V., Juven, P. A., Pontille, D., Saraç-Lesavre, B., and Yon, G. 2017. *Capitalization—A Cultural Guide.* Paris: Presses des Mines.

Muniesa, F., and Linhardt, D. 2011. Trials of Explicitness in the Implementation of Public Management Reform. *Critical Perspectives on Accounting* 22 (6): 550–566.

Nadaï, A., and Labussière, O. 2010. Birds, Turbines and the Making of Wind Power Landscape in South France (Aude). *Landscape Research* 35 (2): 209–233.

Nadaï, A., and Labussière, O. 2017. Exhaustible-Renewable Wind Power. In *The Routledge Research Companion to Energy Geographies*, edited by S. Bouzarovski, M. J. Pasqualetti, and V. Castán Broto. London: Routledge.

Nadaï, A., Labussière, O., Banos, V., Chauvin, C., Cointe, B., Dehez, J., Fontaine, A., Reverdy, T., and Tabourdeau, A. 2018. New Energy Resources in the Making. In *Energy Transitions—A Socio-technical Inquiry*, edited by O. Labussière and A. Nadaï, 49–100. New York: Palgrave Macmillan.

Nadaï, A., Labussière, O., Debourdeau, A., Régnier, Y., Cointe, B., and Dobigny, L. 2014. French Policy Localism: Surfing on "Positive Energie Territories" (Tepos). *Energy Policy* 78: 281–29.

Poize, N. 2015. *Etude du cadre législatif et réglementaire applicable au financement participatif des énergies renouvelables*. ADEME Research Report 1510C0042, Rhônalpénergie-Environnement. Paris: ADEME

Poize, N., and Rüdinger, A. 2014. *Projets citoyens pour la production d'énergie renouvelable: une comparaison France-Allemagne*. IDDRI Working Papers n°01/14, Iddri, Paris.

Simondon, G. 1989. *Du mode d'existence des objets techniques*. Paris: Aubier.

Woolgar, S., and Lezaun, J. 2013. The Wrong Bin Bag: A Turn to Ontology in Science and Technology Studies? *Social Studies of Science* 43 (3): 321–340.

III Turning Nature into Assets

7 Expropriating the Future: Turning Ore Deposits and Legitimate Expectations into Assets

Paul Robert Gilbert

Introduction

In this chapter, I draw on ethnographic fieldwork carried out in the City of London between 2012 and 2014 to explore the valuation techniques that underpin investment in mineral exploration. At the time of my fieldwork, carried out with investors, analysts, consulting geologists, and lawyers, the mining market was seeking out new frontiers as established resource-rich jurisdictions became less profitable, or began to show signs of what analysts termed "resource nationalism" (the desire to renegotiate contracts, taxation, and royalty rates in terms more favorable to host states). Against this background, the chapter explores the capitalization devices, narratives, and durable legal foundations through which mineral deposits are transformed into assets, valued in terms of the future revenue they promise their investors.

The first section of this chapter situates my concern with the assetization of mineral deposits in relation to existing ethnographic work that explores valuation and the political economy of mineral deposits, but which tends to emphasize commodity situations rather than the asset condition. The ethnographic portion of the chapter then moves to examine the socio-technical dimensions of capitalization, with an emphasis on the Real Options-based valuation techniques used to value mineral deposits as assets. Subsequently, I examine the discursive and narrative components of the capitalization process. Specifically, I focus on the manner in which discourses around resource nationalism are deployed by investors in tandem with Real Options-based valuation models to argue against fiscal redistribution in host jurisdictions, and in favor of rapid resource extraction. Finally, I emphasize the increasing levels of protection provided for investors' "legitimate expectations" in international investment law. The chapter therefore argues that it is the assetization of investors' legitimate expectations that future cash flows will materialize, unimpeded by putative acts of resource nationalism, which sustains mineral exploration and investment. These expectations in turn rely upon the techno-economic processes that enable the transformation of largely invisible mineral deposits into assets valued in terms of future revenue streams.

From Minerals as Commodities to Mineral Deposits as Assets

How—and why—do mineral deposits become valuable? Anthropologists studying sapphire (Walsh 2004, 2010), silver (Ferry 2002, 2005), and coltan (Mantz 2008; Smith 2011) extraction have approached this as a question about the value transformations to which minerals are subject as they become *commodities* (see Braun, this volume). Following Arjun Appadurai (1986), this means focusing less on the qualities of given minerals than on the commodity situations through which exchangeability becomes the socially relevant feature of a given metal or gemstone (Appadurai 1986, 13).[1] As such, attention may be given to the value transformations that mineral specimens undergo as they are diverted from circulating within the moral economy of a mining community, to become commodities valued for their beauty and "naturalness" among geographically distant collectors (Ferry 2005, 430; see also Walsh 2010).

Equally of concern for these anthropologists has been the manner in which sites of mineral extraction, and the lives of those involved in mineral commodity situations, are implicated in broader systems of political economy. A great deal of attention is thus given to the manner in which artisanal miners in the Democratic Republic of the Congo (Smith 2011) or Madagascar (Walsh 2004) speculate over the cause of fluctuations in the world prices[2] of the resources that they help transform into valued commodities. While seemingly the product of remote and opaque forces, these price fluctuations have profound implications for miners' capacities to build predictable economic futures for themselves. Hence for Jeffrey Mantz (2008, 41–42) an understanding of the coltan (or "digital mineral") trade demands a perspective that integrates "the seductive virtual world of Halo 3" (and the boom in coltan prices that resulted from the launch of Sony's PlayStation 2), as well as "Citibank and other corporations [who] have negotiated directly with the ruthless occupants of the eastern DRC…who forced people to mine and plundered their villages."

These ethnographies of mineral extraction, resource-making, and commoditization are powerful precisely because they operated on what Jean and John Comaroff term an awkward scale. Such ethnographies, the Comaroffs argue, are equally attentive to "processes occurring in virtual spaces as to those visible in 'real' places-under-production" (Comaroff and Comaroff 2003, 169). Yet, by locating the financial arrangements and capitalist agencies that format mineral commodity chains and the lifeworlds of artisanal miners, such ethnographies render finance a central and yet curiously neglected figure which remains "present, yet largely absent" (Ouma 2015, 163; also Visser 2017). By contrast, the socio-technical arrangements upon which financial agency relies has been central to social studies of finance (SSF). Suspicious of the treatment of finance as a virtual domain (see MacKenzie 2007), SSF scholars ask instead that attention be paid to the endless material and socio-technical work through which goods are marketized, prices calculated, and economic models performed (Callon 1998; Callon and Muniesa 2005; MacKenzie 2001).

SSF has given rise to a thriving literature that traces out the socio-technical arrangements through which securities and derivatives are priced, carbon markets are enacted, and algorithmic trading produces its effects (e.g., Beunza and Garud 2007; Coombs 2016; Lenglet 2011; Lepinay and Callon 2009; MacKenzie 2009; MacKenzie et al. 2012). But while recent work in this tradition has engaged with the relationship between particular models and the 2008 global finance crisis (e.g., MacKenzie and Spears 2014), there has been a tendency to shy away from questions of capital, capitalism, and political economy on the macro scale. Callon (2005; Callon and Latour 1997; also Barry and Slater 2002) has been particularly hostile toward the practice of conjuring capitalism as an explanatory force, and its short-circuiting of the careful tracing through which socio-technical market agencements can be discerned.[3] Defenders of the SSF approach have thus accused the political economists who locate financial agency in elite pacts and capitalist interests (e.g., Froud et al. 2011) of overdetermined analyses resting on conspiratorial thinking (Beunza 2010), and have argued instead that "market institutions and their embedded political controversies are in fact *best* made visible through the description of *devices*" (Lenglet 2011, 46, emphasis added).

If anthropologists occasionally neglect market devices and conjure financial markets as the virtual backdrop against which value transformations play out around particular mineral extraction sites, SSF scholars conducting fieldwork in financial centers have also been guilty of operating with a geographical imagination that is far more circumscribed than that deployed by anthropologists working on an "awkward scale." In spite of Latour's (2005, 179) injunction to focus on neither capitalism nor "the screen of the trading room either," scholars indebted to his material sociology have been deliberately selective in their attention to calculative arrangements within funds' offices, even when those funds trade in the currencies of "far-flung countries" (Hardie and MacKenzie 2009, 66; Beunza et al. 2006). In the terms of critics like Roberts (2012, 42), excessive attention to the "empirical plane of organization" leads to a neglect of "immanent *capitalist* processes." However, by switching attention away from calculative market devices that facilitate the value and exchange of commodities, and toward the capitalization devices (Doganova and Muniesa 2015) through which *assets* are made valuable in light of their earning potential, it is possible to remain attentive to the political controversies embedded in socio-technical devices while also appreciating them as explicitly capitalist forms of calculation that "connect distant situations and configure large social realities" (Muniesa et al. 2017, 17).

The ethnography of capitalization presented below begins from the assumption that "formulating reality in capitalist terms might, in some circumstances, transform the formulated reality" (Muniesa 2014, 35). In other words, the techno-economic operations through which things such as mineral deposits are turned into assets, capitalized as a source of (expected) future revenue, are fundamental to the operation of extractive industry techno-capitalism. The political economists Jonathan Nitzan and Shimshon Bichler have argued forcefully that the study of capitalist realities must begin with the "ritual of capitalization" through which the net present value of an asset is calculated (Nitzan and Bichler 2009, 17–19; Nitzan 1998). This, they argue, is because capitalists themselves approach capital as capitalized earning capacity, and *not* as actual physical and material infrastructure or inventory. For followers of Nitzan and Bichler, what is being capitalized in the calculation of net present value is "the *power* of governments or corporations to shape and reshape the terrain of social reproduction in their favor *relative* to other organizations attempting to do the same thing" (Di Muzio 2012, 371). But attempts to literally "quantify the valuation-qua-capitalization of class power" (Park and Doucette 2016, 545)[4] forego consideration of the careful socio-technical arrangements, narratives, and legal foundations through which the world that is to be capitalized comes to be configured (Muniesa et al. 2017; also Knafo et al. 2014).

In the next section, I examine those socio-technical arrangements—through which mineral deposits become valued in terms of discounted future earnings. I highlight the tensions that arise between miners who conceive of a resources value in terms of sheer tonnage of rock or good community relations, and those who evaluate mineral deposits only in light of their "asset condition" (Muniesa et al. 2017, 32). Specifically, I focus on the models that mineral exploration firms and their investors use to adjust mine plans (including labor contracting, waste management protocols, and the cut-off grade below which extracting rock is seen as unprofitable) with a view to maximizing their net present value.

Afterward, I examine the discursive formations which nourish attempts to capitalize on mineral deposits. To value an asset in light of the future revenue it will produce, investors must have confidence in the *stability* of their mineral claim, and the contractual arrangements that pertain between mineral exploration firms, host states, and mine-area communities. During 2012–2014, mining analysts and investors in the City of London were particularly concerned about the political risk posed by a putatively global tide of "resource nationalism," that was said to be a response to a prolonged mining and metals commodity price boom (or supercycle). The central risk preoccupying explorers, investors, and analysts during this time was the risk of contract review (Gilbert 2020), and the possibility that mining companies may be asked to pay higher royalty and taxation rates on the revenue that their assets produced in response to the commodity price boom. While the putative threat of resource nationalism encouraged some analysts and investors to heavily discount the net present value of their mineral assets, the resource nationalism narrative, in tandem with the capitalization devices used to value and optimize mine plans, has the effect of justifying rapid extraction of a mineral resource, regardless of the consequences for host communities and states.

Finally, I focus on the legal foundations which make the *promise* of capitalization feasible: capitalization devices that render mineral deposits value in terms of their earning potential, and a set of discourses according to which the "viability of all things resides, primarily, in the asset condition" (Muniesa et al. 2017, 52), are not in themselves sufficient to buttress the operations of mineral exploration firms seeking out new resource frontiers.

Investors must be able to point to durable legal configurations that provide confidence in their ability to realize contracted revenue streams in the future. Drawing briefly on a case study from Bangladesh, an emerging extractive industry frontier, I focus on the norms of international investment law and the protection afforded by investor-state arbitration tribunals to extractive industry firms who feel their "legitimate expectations" have been violated by acts of resource nationalism (which may include renegotiated royalty and taxation rates). The Bangladesh case study shows that the norms of investor-state arbitration allow not only mineral deposits but legitimate expectations themselves to be configured as assets. In the conclusion, I reflect on the implications of a highly contagious "capitalization syndrome" (Muniesa et al. 2017) for the ability of host states and communities to pursue alternative models of resource extraction.

Capitalization Devices: Real Options and Mineral Prospects

A concern with the economic ends to which mineral deposits might be put has long animated the discipline of geology (Braun 2000). Geologists are, however, frequently at pains to point out that a mineral deposit is not a resource if it cannot be "brought to book," and that socio-political factors play as much of a part in transforming a mineral deposit into a valuable resource as geology itself (e.g., Edwards and Atkinson 1986; Njowa et al. 2014; Ténière 2015). As Pete, an economic geologist (and officially recognized "Competent Person" authorized to sign off on the disclosures made by exploration firms seeking to list on London's Alternative Investment Market) put it to me in early 2014: "A resource is made up of tonnage and political and economic factors." As such, many geologists and mining analysts preempt through their practice the critique that Richardson and Weszkalnys (2014, 7–15) level at "capitalist forms of resource extraction" which allegedly cast resources as dead matter or inert substances in nature waiting to be discovered. It is certain that consulting economic geologists and mining analysts are aware of the painstaking calculative work and delicate political settlements that must be in place before a resource can be valued and made valuable. Bringing a mineral deposit to book means explicitly bringing it under the capitalizing gaze and coming to appreciate the deposit in terms of the discounted revenue streams it will produce. But this process of transforming invisible—and in the early phases of exploration, merely hypothesized—resources into owned and tradable mineral deposits is first and foremost a techno-economic concern. The generation of contemporary extractive capitalism hinges on the successful deployment of a specific set of calculative practices: mineral deposit classification using specific geostatistical techniques and the valuation of variously classified mineral deposits through discounted cash flow (DCF) or real options (RO) capitalization models.

Here, I outline the technical procedures through which geological expertise is harnessed to the "release of value" (Buchanan 2016, 5) for mineral exploration firms and their investors. Of particular concern in this section is the use of net present value (NPV) calculations—derived either from DCF or RO analysis—to reformat pit plans, the planned life of a mine, and the cut-off grades beyond which extracting ore is not seen as viable. Conflicts arise when the capitalizing gaze—and the maximization of NPV—is prioritized over socio-political and environmental concerns around a given mine site.

To capitalize upon a mineral deposit and turn it into an asset that is valued in terms of discounted future cash flows, an exploration firm must first carry out drilling operations on their mineral property (which will require some form of prospecting or exploration license, depending on the jurisdiction). The samples extracted from the drill-cores are used to map strata and ore grade. Using geostatistical techniques introduced by South African mining engineers in the mid-twentieth century and refined at the École des Mines and Royal School of Mines, a resource block model is developed, where drilling samples are used to estimate the grades within given blocks. The key geostatistical innovation which enables reliable resource block models to be generated—and which underpins the valuation techniques outlined below—is known as kriging, after Danie Krige, a South African engineer whose master's thesis (Krige 1951) introduced a series of refinements to the then-standard practice in the Witwatersrand of determining whether or not it would pay to mine a sampled block (or "selective mining unit") by simply calculating the arithmetic mean of the ore grade recorded across a series of samples. Noting the possible presence of high-grade values in a low-grade block of ore, Krige developed a method for explaining variation in samples in terms of a correlation derived from mapping the distance between each sample in a given block.

For contemporary mining engineers and geostatisticians, determining whether it is economic to mine a given block—or what the ore/waste ratio will be in a given block—remains a primary concern (Assibey-Bonsu and

Krige 1999; Rossi and Deutsch 2013). Since the determination of whether it pays to mine a given block is not based on geostatistical modeling alone but also on a set of economic considerations and forecasts, there is a dynamic relationship between the valuation of a mineral asset and the application of geostatistical techniques to generate information about largely invisible mineral deposits. Indeed, the determination of cut-off grades, the levels at which it is economic to mine a given block, changes as mining proceeds and new information emerges (Peattie and Dimitrakopoulos 2013); this aspect of the mining process appears to have rendered mineral assets especially well-suited to RO-based valuation (see below). Confidence intervals around drill-core samples and geostatistical models of resource blocks also have implications for the classification of deposits under various financial reporting or disclosure codes. As such, it is not unheard of among mining analysts in the City of London for small exploration firms to simply flip mineral properties by carrying out some additional drilling, increasing the statistical likelihood that their deposit is a high-quality resource, and so elevating the mineral asset's NPV. As the convener of the Mining for Bankers training courses at Imperial College puts it, there is "a direct correlation between expenditure on drilling and enhancing the valuation of the project" (Buchanan 2016, 125).

The principal codes through which drilling data is translated into estimates of value (and ultimately NPV) are the Australian Joint Ore Reserves Committee (JORC) reporting code and Canadian NI 43-101 mineral disclosure code. The JORC code, inaugurated in the late 1980s, standardized the criteria for reporting measured, possible, probable, and ore in sight mineral deposits. The NI 43-101 code was introduced by the Canadian Securities Administrators as a direct response to the Bre-X scandal, which involved a speculative bubble built up around a Canadian exploration firm which had fraudulently reported on a 70- to 200-million-ounce gold find in Indonesia (see Tsing 2001). The NI 43-101 code requires that mineral prospects are reported as either resources (broken down by increasing degree of confidence into inferred, indicated, and measured) and reserves (broken down into probable and proven). Due to the legal requirement that the code be used by exploration firms seeking to list on Toronto's TSX-V (home to the largest number of exploration firms in the world), and its use as a de facto standard on the London Stock Exchange and its Alternative Investment Market, the code works as a translation device which reformats drilling data and geostatistical models into estimates of a resource's viability that can be capitalized upon. As to what transforms speculative resources into asset-like reserves, the TSX-V is clear: the provision of infrastructure, initial test-mining, and "legal, environmental, social and governmental factors" (Ténière 2015, 37).

Evidently, turning mineral deposits into assets does not only rely upon geostatistical calculation. Resource or reserve estimates are but one input for the capitalization devices (Doganova and Muniesa 2015) through which mineral deposits come to be valued for their earning power. Also of concern are mineral price forecasts, the costs and availability of infrastructure, the acquisition of appropriate permits, the "social license to operate" and the presence of "political risk." The latter two factors, being less amenable to quantification, are folded into the capitalization calculation through the discount rate (see next section). There is also dispute among mineral economists working as consultants in the extractive industries over the appropriate form of capitalization device to deploy: discounted cash flow (DCF) or real options (RO) models. While some influential mineral economists (Buchanan 2016, 9–34) favor the DCF method, others reject it on the grounds that it utilizes a *single* discount rate, applied to all future cash flows, in order to facilitate a "now or never" analysis of the attractiveness of an asset. To its critics, DCF modeling is not capable of accounting for geological risk or the wide range of uncertainties faced by mineral assets exposed to price, labor, taxation, and royalty rate alterations—but RO analysis is (Samis et al. 2006).

RO analysis, introduced as a tool for making capital allocation decisions by Stewart Myers in the late 1970s and adopted more widely by financial analysts in the late 1990s, allows managers to capitalize on emerging opportunities in a way that DCF—which involves a constant discount factor and a cash flow structure known at the outset—does not (Schulmerich 2010). Although mining executives and analysts were originally slow to adopt RO analysis and many junior miners continue to make decisions based on DCF (Ajak et al. 2018), mining investments do appear particularly amenable to RO analysis—or analysis as if they were a series of financial options contracts. Mining investments are largely irreversible and proceed as a series of options: purchasing a property and a license gives the option to explore, exploration gives the option to develop, and development of a mineral prospect gives the option to extract (Slade 2001). Analyzing them as such allows for decision makers to react to changing market conditions and exercise a series of options, should cost and profit considerations require that this be done (see Schulmerich 2010, 24). These options could include the option to alter cut-off grades, to alter the pathway of progression through a planned pit, or even closing and reopening a mine in response to changing mineral commodity prices and their effect on cut-off grades (Ajak et al. 2018).

RO analysis thus allows each individual source of cash flow (or cost) to be discounted for uncertainty individually. As a result, the likely effects on future cash flow of price rises (or falls), labor cost changes, and the possibility of encountering particularly low (or high) quality ore in a given year can be accounted for. One significant result of the application of RO analysis to mineral asset valuation is the emergence of pit optimization models, which use price forecasts to determine cut-off grades—and thus the organization of a pit, the life of a mine, and the required flexibility of a mine workforce (Asad and Dimitrakopoulos 2012; Dimitrakopoulos et al. 2002; Evatt et al. 2011). Pit optimization models predate the use of RO analysis in mineral economics. Building on the geostatistical block models outlined above, pit optimization consultancies expanded rapidly in the early 1990s when computational advances allowed the easy calculation of an optimal pit outline (Whittle 1990). Optimal pit outlines could be determined as the point at which the value of ore in a given series of blocks (available ore multiplied by ore value), minus the cost of waste removal for those blocks, was maximal. While early pit optimization models used DCF approaches to calculate the dollar value of each feasible pit outline (Whittle 1990, 471), it is perhaps unsurprising that pit optimization has been married to RO analysis since its dissemination among financial analysts in the late 1990s. Not only does information about ore grade itself change as mining proceeds and provides more data points for geostatistical block models, but the scheduling of pit outlines with a view to maximizing a mineral's asset value must by necessity respond to a series of contingencies, from commodity price changes to labor costs, and taxation or royalty rates.

If not explicitly performative, pit optimization models certainly have the capacity to "format" the relationship between extractive industries and host polities (cf. Crosby and Hennebery 2016, 1435). Indeed, as Bowman and Moskowitz (2001) show, the use of RO-based valuation has encouraged executives to take a more experimental approach, and consider proposed capital allocation decisions as a series of options. Below, I illustrate the potential for conflicts to arise when the asset logic of RO models produces recommendations for mine length, cut-off grades, and mine planning that are subordinated to, and arise from, the desire to maximize NPV. To do so, I draw on ethnographic material from seminars run for mining analysts and executives between 2013 and 2015 by a pit optimization modeling company, which I will call ExtractCo.

The modelers at ExtractCo, having convinced heavyweight natural resource fund managers in the City of London to use their RO-based optimization approach to assess possible investments, were now seeking out mining analysts, managers and investors to convert to their approach. As one of their executives put it at a 2014 briefing, "There is a disconnect between what mining investors want and what miners want. Bankers want the 'magic number,' the NPV. Miners don't, they want ounces." For ExtractCo, seeking to persuade others to use their models, mining executives and analysts are too concerned with proxies for productivity like ounces and tonnage of ore, mine length, and the cost of leaving equipment idle—without thinking sufficiently about time. In an effort to convince analysts and executives attending a 2013 training session of the infallibility of the capitalizing gaze, one ExtractCo modeler emphasized the importance of flexible labor and equipment management in order to deal with a brief dip in ore grades as mining might progress through a particular layer: "Don't buy the trucks, hire them. When you don't need them, let them go. You pay extra, but it's going to reduce operating expenditure over ten years."

As discussion turned to a then high-profile (now amended) plan for a "jobless" automated mine on South Africa's platinum belt, some analysts and brokers in attendance expressed concern that this would be viewed as hostile and alienating, given the character of labor politics in the region. The ExtractCo modeler jovially responded that they knew what the optimizer would say: less employment means less cost, means higher NPV—to less than universal approval from the audience. A follow-up question was then asked by a prominent mining broker about the model's capacity to deal with environmental liability, since "only going for the high grade and throwing away the rest of the rock [in an effort to maximize NPV] makes a nasty stockpile." The response from ExtractCo reveals a great deal about the temporal politics of the asset condition:

> Well, "nasty" stockpile, I don't look at it as "nasty"—it's *money*. Environmental restoration is a cost, so is CSR [Corporate Social Responsibility]. If you've got to move a village, if the pit is under a village—and it usually is—the CSR has to be

> paid for, so you have to find cash to move the village. If you have cash, you can do CSR and sustainability, *and* give to shareholders. So get the cash first! Recently an African mining minister said to me, "We want *long term*." We said, "Wait a minute, you got a depleting resource, your roads don't work, hospitals don't work, there is high AIDS—do you want that now or in five to ten years? I think now!" Whether it gets there or not [laughter]—you can't model corruption! So make as much as you can now and *then* invest in CSR or sustainability.

To most in attendance, this was a relatively satisfactory response. The flippant regard shown for generic African mining ministers aside, ExtractCo's response reveals the capitalizing gaze according to which the viability of all things lies in the asset condition (Muniesa et al. 2017). But it also points toward the social, environmental, and distributional conflicts over the course of resource developments that are themselves formatted (and exacerbated) by capitalization devices that place a discount on political uncertainty (see Gilbert 2020), and send their devotees a clear message: cash first, politics later. I return to the temporal politics that surround the distribution of mineral wealth in the conclusion. Next, however, I introduce the discourse of resource nationalism which worked in tandem with capitalization devices and calculations of cut-off grades to present any attempt to raise royalty and taxation rates as harmful to *everyone* implicated in the future of a mineral asset, host communities and states included.

Narrative Plots: Political Risk, Resource Nationalism, and the Supercycle

Turning things into assets rests not only on calculative capitalization devices. It also relies on the operation of a capitalizing gaze that is activated and sustained by a particular narrative plot, according to which valuing objects in terms of their earning capacity becomes sensible (Muniesa et al. 2017). Previously, I outlined the mineral reporting codes, geostatistical methods, and real options (RO) analyses through which mineral deposits come to occupy the asset condition. For a mineral deposit to become a capitalized mineral reserve, valued in terms of its earning potential, certain legal, environmental, social, and governmental factors also need to be in place. Turning a mineral deposit into asset requires convincing investors that these revenue streams will be stable and securely contracted, and that a prospective project still produces a positive NPV even when discounted for political risk.

Between 2012 and 2014, as mineral exploration firms and investors in the City of London sought new extractive opportunities in what one fund manager termed the "farther reaches" (see Gilbert 2019), analysts were particularly concerned about the political risk posed by resource nationalism and its apparent relationship with a commodity price supercycle. The mining market's conception of resource nationalism is expansive, but analysts typically saw the risk of contract review (including renegotiation of taxation and royalty rates) as the preeminent political risk. In other words, political risks are those that interrupt the certainty that an assets capitalized earning capacity will hold forth on its promise. Here, I outline the contours of the resource nationalism narrative prominent in the City of London during 2012–2014. I examine the way that the asset condition itself is used as grounds to challenge a diverse set of regulatory measures (including tighter environmental codes, and raised royalty and taxation rates) as aggressive resource nationalism. Higher taxation and higher royalty rates have the effect of raising costs and raising the cut-off grade in mining projects that have been planned with the RO analysis and pit optimization techniques discussed above.

In their survey of executives' and investors' concerns for 2012–2013, Ernst & Young (2012, 7–14) identified resource nationalism as a key political risk facing mining and metals exploration firms. Their conception of resource nationalism was expansive, incorporating moratoria on investment licenses (Mongolia), plans to tax coal based on market prices rather than volume (China), enforcement of higher royalty payments by an anti-corruption commission (Indonesia), as well as moves to legalize nationalization of mineral assets (South Africa) (see Ernst & Young 2014). This all-encompassing approach to classifying nonbeneficial regulatory acts as resource nationalism was shared by many analysts in the City of London, as well as by international investment lawyers providing advisory services to exploration firms (see Gilbert 2020).

Analysts, investors, and prominent political risk consultants like Ian Bremmer also shared an understanding of resource nationalism as a *global* phenomenon linked to an upswing (or supercycle) in commodities prices, and a desire for resource-rich states to secure a greater share of 'rent' for themselves—even if local "social or political upheaval" conditioned the particular form taken by that resource nationalism (Bremmer and Johnston 2009, 151; see also Gayi and Nkurunziza 2017). Resource nationalism itself is thus depicted as cyclical by many consulting mineral economists, seen to arise in response to long-term economic fluctuations (e.g., Joffé et al. 2009). The notion that the metals and mineral price boom beginning in the late 2000s was a supercycle rather than a normal business cycle is often credited to a Citigroup note (Heap 2005) which attributes the price boom to demand from a growing China. What makes it a putative supercycle is the increasing cost of exploration that prevents the increased capital investment required to restore a putative economic balance (cf. Radetzki 2013, 86)—hence the search for new frontier jurisdictions in the "farther reaches" of mining investors' geographical imaginations.

Because the time from exploration to exploitation may be two decades or more, the renegotiation of taxation and royalty rates up from levels agreed with host governments prior to the supercycle was the primary source of political risk in most resource nationalism narratives (Cuddington and Jerret 2008). The mining market's attention was thus captured throughout 2012–2014 by various reports that claimed to announce the end, peak, or persistence of what analysts depicted as anything from a ten-year to thirty-five-year supercycle.[5] During this period, the resource nationalism narrative was fundamental to the activation of the capitalizing gaze. Rumors that the supercycle was about to end could provide a powerful negotiating position for exploration firms, as one editor of industry news source mining.com put it: "Executives have the power to cherry pick which combination of country and project offers the best returns.... The threat of a cancellation—or long delays—could be a powerful incentive for politicians to offer better terms to companies, executives mutter" (Els 2012).

Challenges to the royalty and taxation rate increases proposed or implemented by a range of resource-rich countries were aided by claims that the supercycle, as a historically unusual fluctuation, was a deviation from the normal, lower prices on which explorers and investors hoped to base tax and royalty calculations. This narrative was put to me explicitly by Colin, a geologist-turned-executive who had established a number of exploration firms in Latin America, the Middle East, and North Africa (as well as consulting on mineral code reform for the World Bank in several of those same jurisdictions). After meeting in 2013 at an event hosted by a professional society for economic geologists (all geologists who sign off on the resource or reserve estimates of listing companies must have professional recognition), Colin invited me to his serviced offices in Mayfair, and subsequently to a Pall Mall club at which he was a member, to explain resource nationalism and the supercycle.

At one of our meetings, Colin related how during the 1990s, "London was the fountain of all knowledge on privatization. All countries were realizing government should never be involved in business. I mean, look at Venezuela!" This treatment of any putative acts of resource nationalism as evidence that a jurisdiction was lined up to be the next Venezuela was a common refrain among analysts during 2012–2014. (see Bremmer and Johnston 2009). Colin then reached for a sheet of paper and began to draw. First, he sketched the axes of a graph, and then a flat horizontal line: the expected annual revenue a mine would produce over its life. Then he superimposed a large parabolic curve toward the end of the time series on the graph. "You see, the private sector brings the cost down. Then the price [of the mineral] goes up, and the government says, 'Wait a minute, you're making a lot of money, and we want to take this off you. Oh well, we will nationalize you.' And of course, it doesn't work, because their objective is to make jobs, and there is no reinvestment. In twenty years, you have to privatize again." As crude as this narrative—of which Colin's rendering is not atypical—may be, it plays an important role in the capitalization process. It is brought into relation with the capitalization devices discussed above primarily through references to cut-off grades.

As noted above, real options (RO) approaches allow for mine plans (including the length of mine, flexibility of labor, and environmental impacts) to be manipulated such that the net present value (NPV) of a mineral asset is maximized, thereby formatting relationships between the extractive industries and host communities through the capitalization device. Mineral prices are, obviously, one of the many sources of revenue that can be discounted and incorporated into RO calculations, with higher prices (alongside the costs of extraction and processing) affecting cut-off grades, and the desirability of maintaining stockpiles to be processed at a later date, in mine plans optimized for NPV (Abubakary et al. 2015; Asad et al. 2016). In an email exchange that took place in 2013 with Simon, an influential mineral economist and consulting geologist in the City of London, I asked about the impact of the supercycle on mineral exploration. The response was that "high government royalties, as advocated by IFC/World Bank, are highly counterproductive, as they are simply costs. They effectively raise cut-off grades, and hence reduce the effective exploitation of, and long-term economic benefit for the host from, the mineral deposit (shorter life, effect on community, etc.)."

The extent to which the World Bank Group advocates higher royalties is certainly subject to contestation, and some brokers and analysts describe the political risk guarantees and co-financing provided by the IFC as a saving grace when attempting to develop projects in high-taxation jurisdictions.[6] However, the email exchange cited above reveals the extent to which the resource nationalism narrative works together with the capitalization device of RO-based mine optimization to challenge the diverse attempts that host jurisdictions make to renegotiate royalty rates. By treating mineral deposits as assets valued for their capacity to produce future revenue, royalty and taxation rates appear necessarily as costs, curtailing the life of a mine, necessitating rapid extraction and potentially undermining host communities' attempts to build themselves a durable economic future.[7] Similarly, assimilating all attempts to renegotiate royalty and taxation rates to an unprecedented supercycle works rhetorically such that it is "through emphasis on excess that the normal is able to emerge" (De Goede 2005, 81). Effectively, the resource nationalism/supercycle narrative allows a range of public policy decisions in diverse resource-rich jurisdictions to be treated as deviations that impede upon the legitimate expectations of those who would transform mineral deposits into assets.

In the final section, I review the relatively recent emergence of this notion of legitimate expectations, and its codification in in international investment law. The norms of international investment law and investor-state arbitration that have emerged over the past four decades form the durable legal foundations that, in the final instance, allow for the assetization of legitimate expectations themselves.

Durable Foundations: Capitalizing on Legitimate Expectations

The resource nationalism/supercycle narrative has the effect of normalizing low royalty and taxation rates negotiated with resource-rich states at the bottom of commodity price cycles. But the cyclical temporality of this narrative, which attributes an internal logic to the oscillations of the market, rests on an erasure of the efforts that postcolonial states made to assert sovereignty over their natural resources during the mid-twentieth century. Equally, it naturalizes the curious and contested emergence of investors' legitimate expectations as a salient legal category that can itself come to inhabit the asset condition. Opposition to previously colonized resource-rich states being beholden to royalty and taxation regimes established by colonizing powers (who tended to treat domestic extractive industry corporations operating in the colonies favorably) was crystallized in the Third World jurists or Third World Approach to International Law (TWAIL) movement in the 1970s and 1980s (Anghie 2007). The arguments put forward by these jurists—that it was entirely unjust for corporations to benefit from concessions or taxation rates agreed with pre-independence administrations when resource prices were historically low—is a direct inversion of the arguments put forward by analysts, mineral economists, and consulting geologists in the contemporary City of London. Where contemporary mining analysts and investors find their rhetorical purpose served by naturalizing *low* points in mineral price cycles, the reverse was true for the Third World jurists' project.

Permanent sovereignty over natural resources was touted as a response to the inequity of colonial concessions, that would allow for renegotiation of taxation and royalty rates—or implementation of a windfall tax—if resource prices boomed (Hossain 1983). Despite UN resolutions recognizing and declaring permanent sovereignty over natural resources during the 1960s and 1970s, the broader project to institute a New International Economic Order of which these efforts were a part crumbled during the 1980s (Bair 2003). At the same time, bilateral investment treaties (BITs), which provided for arbitration between host states (as one party to the treaty) and foreign corporations (domiciled in the state that forms the other party), proliferated, and the number of investor-state arbitrations taking place in the World Bank's International Court for the Settlement of Investment Disputes (ICSID) skyrocketed (Parra 2012). BITs, according to their advocates, provide the confidence required by foreign investors in capital exporting countries, and are thus putatively beneficial to the populace of capital importing countries. Layers operating in the TWAIL tradition, however, view BITs as sustaining an imperial system of investment law, according to which transnational corporations' freedom to contract is given priority over the rights of postcolonial nations, who may be sued for legislative decisions taken in the public interest, or attempts to exert sovereignty over their natural resources (Anghie 2007; Gilbert 2018; Sornarajah 2016).

As more and more investor-state arbitral decisions have been made public, a number of prominent critics have voiced their concern about the expansive interpretation of terms like expropriation, investment, and legitimate expectations (e.g., Prieto-Rios 2015). Of particular concern to critics of international investment law is the extent to which investors may be compensated for loss of future revenue based on the treatment of a regulatory decision as an act of "creeping expropriation." I now want to review the debates that have taken place among scholar-practitioners who work as lawyers or valuation consultants in investor-state arbitrations, regarding the appropriate standards for valuing and compensating for a violation of legitimate expectations. I do this in order to show that a discourse of resource nationalism parallel to that outlined above enables the treatment of legitimate expectations as an asset, such that compensation can be offered for the loss of earnings that might be expected in a favorable investment climate.

In a recent review of the notion of indirect expropriation, international arbitrator Mark Kantor adopts the definition included in the US's Model Bilateral Investment Treaty, whereby indirect expropriation is "the extent to which the government action interferes with distinct, reasonable investment-backed decisions" (Kantor 2015, 173). Other scholar-practitioners define legitimate expectations as the entitlement to protection from harm caused by the retraction of a publicly taken position by an authority, or locate legitimate expectations in the rule of law and the putative necessity for individuals to be able to foresee the consequences of their actions in order to carry out rational capitalist enterprise (Potesta 2013, 7; Schultz 2015; cf. Smith 2011). A particularly troubling area for many arbitral lawyers is the emergent concept of creeping regulatory expropriation whereby no single act constitutes expropriation, but apparently "expropriatory" taxation rates may do so (Kantor 2015, 179). It is precisely the malleable nature of the concept of legitimate expectations that means it has become tribunals' "preferred way of providing protection to claimants in situations where the tests for 'regulatory taking' appear to difficult, complex and too easily assailable" (Potesta 2013, 7). It is perhaps unsurprising that City of London lawyers advising on mineral exploration frequently warned against creeping expropriation as an aspect of resource nationalism. Such creeping expropriation—or impingement on legitimate expectations—could, extractive industry lawyers argued, begin with stringent environmental regulations or the kinds of windfall tax promoted by Third World jurists (see Gilbert 2020).

If a corporation successfully sues a resource nationalist state in an arbitral court for violating its legitimate expectations or for creeping expropriation, compensation may be paid. Compensation for the violation of legitimate expectations rests upon the habit of capitalization but is fraught with difficulty. For instance, if legitimate expectations are violated by a resource nationalist government, how should political risk be incorporated into the valuation of the mineral asset that has been expropriated (and the future earnings that have been lost)? Should firms have already discounted their expected cash flows for a high level of political risk, perhaps as measured by the International Country Risk Guide (Chinen 2016)? Or can (retrospectively) mistaken calculations of NPV based on a *lower* political risk environment be upheld (Joffé et al. 2009; Kantor 2009)? The speculative nature of compensation for expropriation and violation of legitimate expectations continues to trouble arbitrators, and several attempts have been made to develop more rigorous or scientific standards for compensation calculation (Pan 2014; Simmons 2012).[8]

The challenges posed by terms like legitimate expectations and definitions of creeping expropriation exceed these questions of how to refine capitalization devices used in compensation calculations. It is the same narratives about resource nationalism outlined above, and the treatment of supercycles as deviations from normal price levels (thus erasing attempts by Third World jurists to treat light-touch taxation and royalty regimes as the product of artificially low prices) which renders the assetization of mineral deposits durable. Host states' attempts to present their natural resources as destroyed or expropriated national assets have typically failed—a recent attempt is exemplified by Bangladesh suing Canadian exploration firm Niko at ICSID for the future value of gas lost in a blowout putatively due to negligence (see Gilbert 2017). After blowouts took place at one of Niko's concessions in the Chhatak district during 2005, the government of Bangladesh filed suit against them in 2008, demanding compensation for destroyed national assets, and the effective incineration of a portion of the nation's energy future. The legality of the Niko concession had in fact been challenged in court in 2005 by the Bangladesh Environmental Lawyers' Association (BELA) who argued that incorrect classification of the Chhatak field (as a "marginal" resource rather than "abandoned" reserve) had resulted in undue calculation of this asset's earning capacity (and so an excessively lenient royalty and taxation regime). In both cases, the government and BELA invoked Bangladesh's sovereignty over natural resources (enshrined in the constitution), giving credence to Sornarajah's (2016, 1976) claim that the Third World jurists project has not been entirely overwritten.

While attempts to sue resource extraction firms for the destruction of national assets and the erosion of national energy futures have been unsuccessful, attempts by firms like Niko to sue host states for payments withheld as a result of environmental damage they may themselves be liable for have been upheld by ICSID.[9] The extent to which the legitimate expectations of resource extraction firms themselves become assets, underwritten by the durable legal foundations of international investment law, is made even more explicit by yet another ICSID arbitration to which Bangladesh has been party. In the *Saipem v. Bangladesh* (2007) arbitration, arbitrators ruled that the refusal of the Bangladeshi government to pay compensation to a gas company (Saipem) whose project was interrupted by local opposition was itself an act of expropriation. Saipem's legitimate expectations that their capitalized earnings would materialize were buttressed by the conventions of international arbitration despite Bangladeshi courts ruling that they were not eligible for compensation (see Goldhaber 2013). In the final instance, the expectation that future cash flows will materialize, unimpeded by putative acts of resource nationalism and regulatory expropriation, is enabled by norms of international investment law that allow legitimate expectations themselves to be treated as revenue-generating assets.

Conclusion

In this chapter, I have argued that attending to the techno-economic rituals of capitalization allows for an ethnographic perspective that connects political contestation over socio-technical devices to the broader social realities that they configure on a range of awkward geographical scales. The capitalization devices which translate real options valuation of mineral deposits to mine plan optimization models have clear political effects. In the pursuit of the "magic number," NPV, they format labor relations, environmental hazards, and the temporal distribution of revenue among mine-area communities and host states. Several anthropologists have called attention to the disruptive temporal politics that ensue when extractive industry firms attempt to transform mineral deposits into cash up front and do politics later. The ability for mine-area communities to build durable futures is rapidly undermined—just as it is when mine closure plans are constantly deferred when changes to commodity price forecasts alter cut-off grades and render previously worthless blocks valuable (see Gilbert 2012; Gilberthorpe 2008; Kirsch 2014).

I have also argued, after Muniesa et al. (2017), for the significance of discursive formations and narrative plots in sustaining and nourishing the rituals of capitalization. Ducastel and Anseeuw (2017) and Visser (2017) have called attention to the discursive dimensions of land assetization: investors must be convinced of the stability (and scarcity) of land for it to become an asset valued in terms of the revenue it is expected to produce. Similarly, investors in mineral exploration must be convinced of the stability of the contractual arrangements that help to transform mineral deposits into assets, and the absence of political risks including resource nationalism. Any increase in royalty and taxation rates is assimilated to a discourse of resource nationalism, and presented as deviant behavior that simply responds to a rare and artificial mining price supercycle. As I have shown, resource nationalism narratives can be paired with the capitalization device to argue that increasing royalty and taxation rates is *bad* for host communities and states. Real options valuation and pit optimization models can be used to undergird a rhetorical ploy whereby increasing royalty and taxation rates is seen to force miners to shorten the life of a mine—in contravention of technocratic guidelines for the long-term management of resource revenue by states in the Global South (Humphreys et al. 2007).

To this extent, the capitalization device wrests efforts to determine the optimal rate of resource extraction away from indigenous communities (Kirsch 2014), social movements, and social-democratic states (McNeish and Logan 2012). But it would be a mistake to locate the politics of mineral assets in terms of calculative or capitalization devices alone. The "capitalization syndrome" outlined by Muniesa et al. (2017) is able to exert its influence on mineral exploration—and conflicts between mineral extraction companies and host states or communities—in large part because of the durable legal foundations through which the promise of future revenue streams is made to appear feasible. It is the norms of international arbitration, together with the discourse of resource nationalism and political risk, that allow mineral exploration firms and their investors to capitalize on legitimate expectations. It is the putative legitimacy of these expectations that must be punctured if host communities and states are to regain control of the temporal politics of mineral asset management.

Notes

1. Ferry (2005) situates her work in relation to Nancy Munn's (1977) approach to value transformation, on which Appadurai (1986) also drew.

2. On the idea of a "world" price (for cotton), and the many transformations and derivations through which it is produced, see Çalişkan (2010).

3. It is certainly true that "capital" and "capitalism" can figure as a causal or explanatory force that is prominent and yet analytically absent—or at least under-theorized—in much anthropological work on mineral extraction (e.g., Sawyer 2002; also Smith 2011; Walsh 2004). This is not to say that anthropologists are guilty of the totalizing abandonments of agency toward which Callon and Beunza direct their critiques. For instance, James Ferguson's (2005, 379) observation that capital does not "flow" from financial centers to distant sites of extraction, but "hops" between appropriately configured "mineral-extraction enclaves" has been highly influential (e.g., Gardner 2012; Welker 2014).

4. By measuring a rising capital share of income and "non-negative rate of differential accumulation by the dominant capital group" (Park and Doucette 2016, 547).

5. The following examples are all drawn from www.mining.com, a leading industry news site: Anthony Hallie, March 7, 2013, "Commodities 'supercycle' will last another 15 years: JP Morgan"—a view taken on the basis that India will need to build 900 million square meters of residential and commercial space and 400 kilometers of subway annually to account for fifteen forecasted years of urbanization (http://www.mining.com/commodities-supercycle-will-last-another-15-years-jp-morgan-85593/); Cecilia Jasmamie, May 12, 2013, "Commodity supercycle is essentially over: Citigroup"—a claim based on growth slowing in China (http://www.mining.com/commodity-supercycle-is-essentially-over-citigroup-73028/); Anna Komnenic, September 26, 2013, "Rumours of supercycle's death are greatly exaggerated: McKinsey"—based on the idea that future demand from China and, intriguingly, the potential future incorporation of environmental costs into metal prices (http://www.mining.com/rumours-of-the-supercycles-death-are-greatly-exaggerated-mckinsey-48435/).

6. One broker based in a prominent City of London mining investment and advisory firm (interviewed in 2012) explained resource nationalism to me in terms of the difference between Burkina Faso with 25 percent taxation and 2 percent royalty, and Ethiopia with "something like 35 percent and 8 percent." In the case of the latter it is "just *impossible*, unless you bring in the World Bank and the IFC. They look after the projects."

7. Though the assessment of who benefits from rapid extraction made by Simon and the ExtracCo modelers may differ, they do share the same commitment to the capitalizing gaze. For an analysis of the impact of royalties on cut-off grades where

royalties are treated as costs (but taxation is not, because cut-off grades are determined on the basis of pre-tax profits) see Lilford (2017).

8. It should be noted that most discussions of valuation in relation to the protection of legitimate expectations do in fact use DCF models, rather than Real Options analysis—perhaps surprising given the focus of Real Options analysis on the ability to respond to emerging contingencies.

9. See the ICSID cases ARB/10/11 and ARB/10/18.

References

Abubakary, S., Nehring, M., and Greberg, J. 2015. Evaluation of the Impact of Commodity Price Change on Mine Plan of Underground Mining. *International Journal of Mining Science and Technology* 25 (3): 375–382.

Ajak, A. D., Lilford, E., and Topal, E. 2018. Valuing the Unknown: Could the Real Options Have Redeemed the Ailing Western Australian Junior Iron Ore Operations in 2013–2016 Iron Price Crash? *International Journal of Mining, Reclamation and Environment* 33 (8): 1–24.

Anghie, A. 2007. *Imperialism, Sovereignty and the Making of International Law*. Cambridge: Cambridge University Press.

Appadurai, A. 1986. Introduction: Commodities and the Politics of Value. In *The Social Life of Things: Commodities in Cultural Perspective,* edited by Arjun Appadurai, 3–63. Cambridge: Cambridge University Press.

Asad, M., and Dimitrakopoulos, R. 2012. Optimal Production Scale of Open Pit Mining Operations with Uncertain Metal Supply and Long-Term Stockpiles. *Resources Policy* 37 (March): 81–89.

Asad, M., Qureshi, M., and Hyongdoo, J. 2016. A Review of Cut-Off Grade Policy Models for Open Pit Mining Operations. *Resources Policy* 49 (September): 142–152.

Assibey-Bonsu, W., and Krige, D. 1999. Practical Problems in the Estimate of Recoverable Reserves When Using Kriging or Simulation Techniques. *International Symposium on Geostatistical Simulation in Mining,* Perth Australia, October.

Bair, J. 2003. Taking Aim at the New International Economic Order. In *The Road from Mont Pélerin: The Making of the Neoliberal Thought Collective,* edited by P. Mirowski and D. Plehwe, 347–385. Cambridge, MA: Harvard University Press.

Barry, A., and Slater, D. 2002. Technology, Politics and the Market: An Interview with Michel Callon. *Economy and Society* 31 (2): 285–306.

Beunza, D. 2010. Political Economists Denounce Social Studies of Finance for Overlooking the Political, https://socfinance.wordpress.com/2010/05/21/political-economists-denounce-social-studies-of-finance-for-overlooking-the-political/.

Beunza, D., and Garud, R. 2007. Calculators, Lemmings or Frame-Makers? The Intermediary Role of Securities Analysts. *Sociological Review* 55 (S2): 13–39.

Beunza, D., Hardie, I., and MacKenzie, D. 2006. A Price Is a Social Thing: Towards a Material Sociology of Arbitrage. *Organization Studies* 27 (5): 721–745.

Bowman, E., and Moskowitz, G. 2001. Real Options Analysis and Strategic Decision Making. *Organization Science* 12 (6): 772–777.

Braun, B. 2000. Producing Vertical Territory: Geology and Governmentality in Late Victorian Canada. *Ecumene* 7 (1): 7–46.

Bremmer, I., and Johnston, I. 2009. The Rise and Fall of Resource Nationalism. *Survival* 51 (2): 149–158.

Buchanan, D. 2016. *Metals and Energy Finance: Advanced Textbook on the Evaluation of Mineral and Energy Projects*. London: Imperial College Press.

Çalişkan, K. 2010. *Market Threads: How Cotton Farmers and Traders Create a Global Commodity*. Princeton, NJ: Princeton University Press.

Callon, M. 1998. Introduction: The Embeddedness of Economic Markets in Economics. *Sociological Review* 46 (S1): 1–57.

Callon, M. 2005. Why Virtualism Paves the Way to Political Impotence: A Reply to Daniel Miller's Critique of *The Laws of the Market*. *Economic Sociology* 6 (2): 3–20.

Callon, M., and Latour, B. 1997. Tu ne calculeras pas!—ou comment symétriser le done et le capital. *Revue du MAUSS* 9: 45–70.

Callon, M., Méadel, C., and Rabeharisoa, V. 2002. The Economy of Qualities. *Economy and Society* 31 (2): 194–217.

Callon, M., and Muniesa, F. 2005. Economic Markets as Calculative Collective Devices. *Organization Studies* 26 (8) 1229–1250.

Chinen, M. 2016. The Standard of Compensation for Takings. *Minnesota Journal of International Law* 25 (2): 335–380.

Comaroff, J., and Comaroff, J. 2003. Ethnography of an Awkward Scale: Postcolonial Anthropology and the Violence of Abstraction. *Ethnography* 4 (2): 147–179.

Coombs, N. 2016. What's an Algorithm? Financial Regulation in the Era of High-Frequency Trading. *Economy and Society* 45 (2): 278–301.

Crosby, N., and Hennebery, J. 2016. Financialization, the Valuation of Investment Property and the Urban Built Environment in the UK. *Urban Studies* 53 (7): 1424–1441.

Cuddington, J., and Jerett, D. 2008. Super Cycles in Real Metals Prices? *IMF Staff Papers* 55 (4): 541–565.

De Goede, M. 2005. *Virtue, Fortune, and Faith: A Genealogy of Finance*. Minneapolis: University of Minnesota Press.

Dimitrakopoulos, R., Farrelly, C., and Godoy, M. 2002. Moving Forward from Traditional Optimization: Grade Uncertainty and Risk Effects in Open-Pit Design. *Transactions of the Institution of Mining and Metallurgy*, 111 (1): 82–88.

Doganova, L., and Muniesa, F. 2015 Capitalization Devices: Business Models and the Renewal of Markets. In *Making Things Valuable,* edited by M. Kornberger, L. Justesen, A. K. Madsen, and J. Mouritsen, 109–125. Oxford: Oxford University Press.

Ducastel, A., and Anseeuw, W. 2017. Agriculture as an Asset Class: Reshaping the South African Farming Sector. *Agriculture and Human Values* 34 (1): 199–209.

Els, F. 2012. End of the Supercycle Is Also Spelling the End of Radical Resource Nationalism, http://www.mining.com/end-of-the-supercycle-is-also-spelling-the-end-of-radical-resource-nationalism/.

Ernst & Young. 2012. *Business Risks Facing Mining & Metals 2012–2013*. London: Ernst & Young.

Ernst & Young. 2014. *Mining & Metals—Resource Nationalism Update*. London: Ernst & Young.

Evatt, G., Johnson, P., Morarity, J., Duck, P., Howell, S., and Tonkin, C. 2011. The Resource Valuation and Optimisation Model—Real Impact from Real Options. *353rd APCOM Symposium, Wollongong,* September 30.

Ferguson, J. 2005. Seeing like an Oil Company: Space, Security and Global Capital in Neoliberal Africa. *American Anthropologist* 107 (3): 377–382.

Ferry, E. 2002. Inalienable Commodities: The Production and Circulation of Silver and Patrimony in a Mexican Mining Cooperative. *Cultural Anthropology* 17 (3): 331–358.

Ferry, E. 2005. Geologies of Power: Value Transformations and Mexican Mineral Specimens. *American Ethnologist* 32 (3): 420–436.

Froud, J., Moran, M., Milsson, A., and Williams, K. 2011. Opportunity Lost: Mystification, Elite Politics and Financial Reform in the UK. *Socialist Register* 47: 98–119.

Gardner, K. 2012. *Discordant Development: Global Capitalism and the Struggle for Connection in Bangladesh*. London: Pluto.

Gayi, S., and Nkurunziza, J. 2017. Trends in Minerals, Ores and Metals Prices. In *The Political Economy of Natural Resources and Development: From Neoliberalism to Resource Nationalism,* edited by P. Haslam and P. Heidrich, 35–52. London: Routledge.

Gilbert, P. 2012. The Development of Difference: Social Change around the Ok Tedi Copper and Gold Mine, Papua New Guinea. *Durham Anthropology Journal* 18 (1): 61–114.

Gilbert, P. 2017. Social Blowouts: Practicing Comparison and Locating Expertise in Bangladesh's Extractive Industries. *Paper Presented at the London School of Economics and Political Science, Department of Anthropology*, March 10.

Gilbert, P. 2018. Sovereignty and Tragedy in Contemporary Critiques of Investor-State Dispute Settlement. *London Review of International Law* 6 (2): 211–231.

Gilbert, P. 2020. Speculating on Sovereignty: "Money Mining" and Corporate Foreign Policy at the Extractive Industry Frontier. *Economy and Society* 1 (49), https://doi.org/10.1080/03085147.2019.1690255.

Gilberthorpe, E. 2014. The Money Rain Phenomenon: Papua New Guinea Oil and the Resource Curse. In *Natural Resource Extraction and Indigenous Livelihoods: Development Challenges in an Era of Globalization,* edited by E. Gilberthorpe and G. Hilson, 75–92. Farnham, UK: Ashgate.

Goldhaber, M. D. 2013. The Rise of Arbitral Power over Domestic Courts. *Stanford Journal of Complex Litigation* 1 (2): 373–416.

Hardie, I., and MacKenzie, D. 2007. Assembling an Economic Actor: The Agencement of a Hedge Fund. *Sociological Review* 55 (1): 57–80.

Heap, A. 2005. China—The Engine of a Commodities Super Cycle. *Citigroup Global Markets, Paper Presented at Mineral Economics and Management Society Annual Conference*, Washington, April.

Hossain, K. 1983. Introduction. In *Permanent Sovereignty over Natural Resources in International Law,* edited by K. Hossain and S. R. Chowdhury, ix–xx. London: Frances Pinter.

Humphreys, M., Sachs, J., and Stiglitz, J. 2007. Future Directions for the Management of Natural Resources. In *Escaping the Resource Curse,* edited by M. Humphreys, J. Sachs, and J. Stiglitz, 322–336. New York: Columbia University Press.

Joffé, G., Stevens, P., George, T., Lux, J., and Searle, C. 2009. Expropriation of Oil and Gas Investments: Historical, Legal and Economic Perspectives in a New Age of Resource Nationalism. *Journal of World Energy Law and Business* 2 (1): 3–23.

Kantor, M. 2009. New Age Compensation for Expropriation. *Journal of World Energy Law and Business* 2 (2): 155–162.

Kantor, M. 2015. Indirect Expropriation and Political Risk Insurance for Energy Projects. *Journal of World Energy Law and Business* 8 (2): 173–198.

Kirsch, S. 2014. *Mining Capitalism: The Relationship between Corporations and Their Critics.* Berkeley: University of California Press.

Knafo, S., Hughes, M., and Wyn-Jones, S. 2014. Differential Accumulation and the Political Economy of Power. In *The Capitalist Mode of Power: Critical Engagements with the Power Theory of Value,* edited by T. Di Muzio, 134–151. London: Routledge.

Krige, D. 1951. A Statistical Approach to Some Basic Mine Valuation Problems on the Witwatersrand. *Journal of the Chemical, Metallurgical and Mining Society of South Africa* 52 (6): 119–139.

Latour, B. 2005. *Reassembling the Social: An Introduction to Actor–Network–Theory*. Oxford: Oxford University Press.

Lenglet, M. 2011. Conflicting Codes and Codings. *Theory, Culture & Society* 28 (6): 44–66.

Lepinay, V., and Callon, M. 2009. Sketch of Derivations in Wall Street and Atlantic Africa. In *Accounting, Organizations and Institutions: Essays in Honour of Anthony Hopwood*, edited by C. Chapman, D. Cooper, and P. Miller, 259–289. Oxford: Oxford University Press.

Lilford, E. 2017. Quantitative Impacts of Royalties on Mineral Projects. *Resources Policy* 53: 369–377.

MacKenzie, D. 2001. Physics and Finance: S-Terms and Modern Finance as a Topic for Science Studies. *Science, Technology & Human Values* 26 (2): 115–144.

MacKenzie, D. 2007. The Material Production of Virtuality: Innovation, Cultural Geography and Facticity in Derivatives Markets. *Economy and Society* 36 (3): 355–376.

MacKenzie, D. 2009. *Material Markets: How Economic Agents Are Constructed*. Oxford: Oxford University Press.

MacKenzie, D., Beunza, D., Millo, Y., and Pardo-Guerra, J. 2012. Drilling through the Allegheny Mountains: Liquidity, Materiality and High-Frequency Trading. *Journal of Cultural Economy* 5 (3): 279–296.

MacKenzie, D., and Spears, T. 2014. The Formula that Killed Wall Street: The Guassian Copula and Modelling Practices in Investment Banking. *Social Studies of Science* 44 (3): 393–417.

Mantz, J. 2008. Improvisational Economies: Coltan Production in the Eastern Congo. *Social Anthropology* 16 (1): 34–50.

Mbembé, A. 2003. Necropolitics. *Public Culture* 15 (1): 11–40.

Muniesa, F., Doganova, L., Ortiz, H., Pina-Stranger, Á., Paterson, F., Bourgoin, A., Ehrenstein, V., Juven, P., Pontille, D., Saraç-Lesavre, B., and Yon, G. 2017. *Capitalization: A Cultural Guide*. Paris: Mines Paris Tech,

Munn, N. 1977. The Spatiotemporal Transformations of Gawan Canoes. *Journal de la Société des Oceanistes* 33 (54/55): 39–53.

Nitzan, J. 1998. Differential Accumulation: Towards a New Political Economy of Capital. *Review of International Political Economy* 5 (2): 169–216.

Nitzan, J., and Bichler, S. 2009. *Capital as Power: A Study of Order and Creorder*. London: Routledge.

Njowa, G., Clay, A., and Musingwini, C. 2014. A Perspective on Global Harmonization of Major National Mineral Asset Valuation Codes. *Resources Policy* 39 (March): 1–14.

Ouma, S. 2015. *Assembling Export Markets: The Making and Unmaking of Global Food Connections in West Africa*. Oxford: Wiley Blackwell.

Pan, J. 2014. Valuation Standards for Calculating ICSID Awards. *Pepperdine Dispute Resolution Law Journal* 14 (3): 355–374.

Park, H., and Doucette, J. 2016. Financialization or Capitalization? Debating Capitalist Power in South Korea in the Context of Neoliberal Globalization. *Capital & Class* 40 (3): 533–554.

Parra, A. 2012. *The History of ICSID*. Oxford: Oxford University Press.

Peattie, R., and Dimitrakopoulos, R. 2013. Forecasting Recoverable Ore Reserves and Their Uncertainty at Morila Gold Deposit, Mali: An Efficient Simulation Approach and Future Grade Control Drilling. *Mathematical Geosciences* 45 (8): 1005–1020.

Potesta, M. 2013. Legitimate Expectations in Investment Treaty Law: Understanding the Roots and the Limits of a Controversial Concept. *ICSID Review* 28 (1): 88–122.

Prieto-Rios, E. 2015. Neoliberal Market Rationality: The Driver of International Investment Law. *Birkbeck Law Review* 3 (1): 55–76.

Radetzki, M. 2006. The Anatomy of Three Commodity Booms. *Resources Policy* 31 (1): 56–64.

Radetzki, M. 2013. The Perseverance of the Ongoing Metal and Mineral Boom. *Mineral Economics* 25 (2-3): 83–88.

Richardson, T., and Weszkalnys, G. 2014. Introduction: Resource Materialities. *Anthropological Quarterly* 87 (1): 5–30.

Roberts, J. 2012. Poststructuralism against Poststructuralism: Actor-Network Theory, Organizations and Economic Markets. *European Journal of Social Theory* 15 (1): 35–53.

Samis, M, Davis, G., Laughton, D., and Poulin, R. 2006. Valuing Uncertain Asset Cash Flows When There Are No Options: A Real Options Approach. *Resources Policy* 30 (4): 285–298.

Sawyer, S. 2002. *Bobbitizing* Texaco: Dis-membering Corporate Capital and Re-membering the Nation in Ecuador. *Cultural Anthropology* 17 (2): 150–180.

Schulmerich, M. 2010. *Real Options Valuation: The Importance of Interest Rate Modelling in Theory and Practice*. Berlin: Springer-Verlag.

Schultz, Thomas. 2015. *Transnational Legality: Stateless Law and International Arbitration*. Oxford: Oxford University Press.

Simmons, J. 2012. Valuation in Investor-State Arbitration: Toward a More Exact Science. *Berkeley Journal of International Law* 30 (1): 196.

Slade, M. 2001. Valuing Managerial Flexibility: An Application of Real-Option Theory to Mining Investments. *Journal of Environmental Economics and Management* 41 (2): 193–233.

Smith, J. 2011. Tantalus in the Digital Age: Coltan Ore, Temporal Dispossession and "Movement" in the Eastern Democratic Republic of the Congo. *American Ethnologist* 38 (1): 17–35.

Sornarajah, M. 2016. On Fighting for Global Justice: The Role of a Third World International Lawyer. *Third World Quarterly* 37 (11): 1972–1989.

Ténière, P. 2015. Technical Disclosure Best Practices and Useful Guidance for Exploration and Mining Companies. *Paper Presented at PDAC*, Toronto, March 4.

Tsing, A. 2000. Inside the Economy of Appearances. *Public Culture* 12 (1): 115–144.

Visser, O. 2017. Running Out of Farmland? Investment Discourses, Unstable Land Values and the Sluggishness of Asset Making. *Agriculture and Human Values* 34 (1): 185–198.

Walsh, A. 2004. In the Wake of Things: Speculating in and about Sapphires in Northern Madagascar. *American Anthropologist* 106 (2): 225–237.

Walsh, A. 2010. The Commodification of Fetishes: Telling the Difference between Natural and Synthetic Sapphires. *American Ethnologist* 37 (1): 98–114.

Welker, M. 2014. *Enacting the Corporation: An American Mining Firm in Post-Authoritarian Indonesia*. Berkeley: University of California Press.

Whittle, J. 1990. Open Pit Optimization. In *Surface Mining*, edited by Bruce Kennedy, 470–477. Littleton, CO: Society for Mining, Metallurgy and Exploration.

8 From Commodity to Asset and Back Again: Property in the Capitalism of Varieties

Veit Braun

Introduction

What does it mean to turn something into an asset—not just conceptually, but also in terms of the good itself? And what are the implications for doing business with others through assets? These are crucial questions as they connect the nature of a good to the issue of what role we want business and economic matters to play in our lives. If Birch (2017a) is correct in his observation that certain industries like biotechnology have abandoned the market and its principles in favor of capitalizing on their assets instead of selling them, and if Muniesa et al. (2017) are right about connecting assets to what is commonly called "capitalism" (as opposed to "market economy"), we need to inquire into the relationship between the nature of a good, including its economic and wider social effects. But how to approach things as elusive as assets?

Birch (2017a, 462) argues that (contrary to many STS approaches to economic matters) the answer is best sought not in the materiality of goods but in the way these goods are valuated through specific forms of calculation and assessment. This is a compelling argument: after all, Tesla's net income from selling electric cars has been consistently negative until recently while Amazon barely makes any profits from its online sales, yet both companies continue to soar at the stock markets, being valued at several billion USD each. If we want to understand the gap between the market value of companies and the market value of the goods they produce, it is not enough to simply look at the latter, Birch rightly concludes—we rather need to look at how they are produced as something valuable. This conclusion, however, also comes with several drawbacks. Firstly, it poses the danger of falsely reifying commodities: who's to say that a commodity's market value is any more abstract, concrete, simple, or complex than that of an asset? Secondly, it solely defines an asset in opposition to a commodity, neglecting other possible contrasts (like commodity–gift or asset–liability). Thirdly, by excluding materiality from the outset, we might deprive ourselves of anything to be found there in spite of Birch's confidence that there will not be all that much. In any case, the argument does not locate the asset in a good itself but outside, in the calculations and conventions of calculative devices and agencies (Callon and Muniesa 2005; Hardie and MacKenzie 2007). In somewhat exaggerated terms, such assets are socially but not materially constructed.

Slightly deviating from Birch's agenda while at the same time taking seriously his call and arguments, I would like to propose a supplementary approach. My interest lies in the question of whether anything can be turned into an asset. If so, could it also be a commodity? If the answer is no, are the reasons to be sought merely in "trials of strength" (Latour 1993) between agents and agencies or does the good itself also have a say in its assetization? To answer this question, I will turn to a controversy between German wheat breeders and farmers over the nature of the seed sold from the former to the latter: should it be treated as a commodity or as an asset—and if so, whose?[1] The answers offered by the involved parties do not only imply different forms of valuation, but also diverging practices of using and shaping the good in question—wheat varieties—in very material ways. What I will be looking at are their specific articulations in order to understand the relationship between assetization and commodification in wheat.

Commodified Seed

For roughly a century now, wheat breeding in Germany has been the business of private companies. Wheat is bred commercially by medium-sized to large firms, many of which are family businesses or part of cooperatives. Unlike in the US (Brandl and Glenna 2017), Australia (Head et al. 2012), or other countries, public wheat breeding has long been confined to basic research and resistance breeding in Germany while breeding by farmers themselves is virtually nonexistent (Harwood 2012; Brandl 2017). Farmers receive breeding companies' seed through commercial distributors (who usually also provide them with other farm input) at around €52 a quintal, the breeder's share (i.e., their income stream) in the final price making up €7.00–13.25, or 13 to 25 percent (2017 prices; STV 2018).[2] Wheat seed is produced for and distributed via anonymous markets: breeders do not know their customers in advance, nor are they aware of the latter's growing conditions. Through experimental crossing and subsequent selection, new varieties of wheat are developed and then evaluated on a large scale—first by the breeders themselves to gather data, at a later stage also by public assessment programs on a national and regional level which provide essential information for comparing and ranking varieties to breeders, farmers, and their respective contractors (Pallauf 2018). A finished plant variety will produce uniform plants which can be clearly distinguished from other varieties and whose agronomic performance will hardly differ over subsequent years. This is essential for farmers who will have to harvest all of their produce within a couple of days and for whom predictable and homogeneous quality matters.

What at first sight looks very much like a model for a market economy nevertheless comes with some major restrictions: only sufficiently standardized wheat varieties which outperform existing ones will be admitted to the market. Without approval by the Federal Variety Office (Bundessortenamt, BSA), breeders are not allowed to sell their seeds to farmers. This regulation is meant to safeguard steady progress in seeds as well as a neat market with no more varieties than farmers could realistically compare with each other. In return, the state grants breeders property rights for their varieties that basically amount to a temporary sales monopoly for twenty-five years. During that time, no one else is allowed to sell seeds of that variety on the market without the breeder's permission. This so-called plant variety protection (PVP) legislation is a unique intellectual property right for plant seeds. While it can also be found in many other member countries of the International Union for the Protection of New Varieties of Plants (UPOV), the legislations of Germany and other European countries exclude plant varieties from patentability, making PVP the only legal instrument to govern intellectual property in this field.

In Germany, seeds may legally only be commercialized in the form of varieties. This means that breeders have to "bring them into shape" before they can take them to the market. If varieties fulfill the criteria of homogeneity (uniform appearance), distinctiveness (phenotypic distinguishability), and stability (stable characteristics over several generation), PVP will legally protect them against "plagiarization" by third parties. This effectively forces breeders to commodify their products, shaping seed in a way that allows market actors to make economically informed, rational decisions (Callon 1999). While some organic farmers and eco-activists openly oppose PVP as a threat to biodiversity and as catering to the needs of industrial agriculture (Aistara 2014; Demeulenaere 2014), it enables transregional seed markets of buyers and sellers who remain anonymous to each other. These markets are not only enabled by distributive and calculative infrastructures like agri-trade firms or technologically enhanced capacities for economic decisions: the strongly standardized nature of the plants themselves is a necessary condition for comparing and evaluating varieties, be it for assessment programs or individual farmers. Bred for context-independence and made "immutable" (Latour 1987), varieties largely work beyond the specific requirements of local soils, climates, and ecologies. Farmers do not need firsthand experience of a variety to assess its performance but can make up their mind by looking at numbers referring to quantifiable, distinguishable products.

That wheat seeds come as standardized, homogeneous, comparable, quantified, and stable objects traded on a market thus gives good reason to characterize them as commodities. As such, they can bridge the geographical, temporal, and social distance between sellers and buyers as well as their respective contexts (cf. Callon 1998). This requires that wheat seeds are thoroughly bred for several generations, evaluated with statistical methods across various climatic regions, and coated with chemicals that protect them from insects and fungi. All these activities are part of a process of commodification, not just semantically, but also materially turning seeds into things that can be sold on an anonymous market. That commodities function so smoothly in combination with markets allows buyers and sellers to be quits (Kopytoff 1986; Callon 1999; Callon and Latour 2011), as they require no complex dis- and re-entanglement, work without assistance or aftercare of the seller (Schubert et al. 2011), and thus give no reason for a prolonged economic or social exchange between transaction partners.

Commodities not only need to be alienable, as Don Slater (2002) has pointed out, they also need to be reappropriable without much effort. This is more than just a matter of access to intelligible and comparable quantified data such as prices, ingredients, and weights. It also involves a material commodification process without which they could not even be quantified in the first place. Commodities also have a social effect beyond alienation between producer and good or seller and buyer. Whereas in former times, farmer and breeder were one and the same person, they are now consumers and producers of seeds, respectively. There is a division of labor along the seed value chain in which farmers turn the value produced by breeders into something else—grain—which in turn is ground to flour by millers a little further downstream, with a hungry consumer waiting at the end of the value chain to finally turn a loaf of bread into energy. In a classical production logic, these are instances of consumption (Graeber 2011): an initial stage of the good is fully exhausted, giving way to a new state of the product, which in turn will be fully used to bring forth yet another form of it. This material zero-sum game is mirrored in the alienation of property in a sales transaction: here, property in a good is fully exhausted on the one side of the transaction (Perzanowski and Schultz 2018, 25–28) and compensated by an equivalent flow of money in the opposite direction (Demsetz 1967; see also Kang, this volume). But if sellers and buyers do not stick to these zero-sum games, the state of seeds as a commodity becomes fragile.

Assets in the Wrong Hands

It is in the nature of seeds that they are able to multiply. A wheat seed sown in fall will yield fifty to sixty seeds in the next summer, provided it is adequately cared for. For breeders, this is a great advantage: they can start with a single cross between two parent plants to create a new variety that quickly scales up to thousands and millions of seeds within few generations. But it is also a threat to their business, as farmers may save a fraction of their harvest for resowing it—instead of buying new seeds from the breeder who initially created the variety. The product is then not fully exhausted in the production process; instead, a part of it remains and multiplies, expanding the material value of a wheat variety—but diminishing its economic worth on the breeder's side. Here we find one mechanism of accumulation (Birch and Muniesa, this volume): a fraction of the harvest does not follow the zero-sum game of consumption but the surplus laws of reproduction. In farming practice, the two differ only in a few details. In both cases farmers harvest their wheat fields mechanically and ship their grain to processing facilities. While in the case of commodification (i.e., consumption), these will be mills, breweries, or fodder producers, assetized (i.e., reproduced) seed ends up in seed treatment plants to be cleaned and chemically coated to be sown again. Both processes turn the harvest into objects of value. Valuation is not just a mathematical procedure here, though, but even more so a biological one—the material contrast between flour and wheat seeds could not be more pronounced. Milled grain continues to flow down the value chain as a commodity; treated seed is held on to and continues to generate value on the farm. Ironically, it is seeds' commodified form (more precisely their standardized, immutable and context-independent nature) which makes it possible to multiply modern wheat seeds without much effort—and thereby to stray from the path of commodification set out for them by their breeders.

PVP, understanding seeds as something akin to intellectual property, was initially created to prevent "horizontal" plagiarism. Breeders were meant to be protected against competitors who would otherwise take their seed, multiply and sell it without having to bear the initial costs for creating a new variety (Sanderson 2017, 21–44). Historically, it therefore only gave a monopoly on commercialization to the breeder. Farmers consented to this restriction, primarily because they lacked the capacities for transregional distribution and sale anyway. For the longest time, the PVP framework in Germany and elsewhere thus implicitly granted a so-called farmers' exemption to them which allowed for saving and reusing seeds on-farm, as long as they did not reenter the market as seeds. This practice, however, became the source of a heated controversy between farmers and breeders from the 1990s on when EU laws rendered seed saving and on-farm reuse illegal if exercised on farms exceeding a few hectares.

In 1991, an amended version of the international UPOV convention on plant variety protection was passed and subsequently implemented in European law in 1994 and German legislation in 1997. Several developments coincided at that time: the wheat market had become more competitive due to increased breeding efforts and varieties lasted shorter on the market but longer on the fields, as phenotypic stability had increased. In earlier times, some varieties would bring in revenues for thirty years (interview with wheat breeder April 2015; Pallauf 2018). Today, farmers are buying less and less new wheat seed on the market (Kempf 2016). Meanwhile, modernization of agriculture had brought about a change of the economic landscape. In this sense, the motives for the move toward assetization of the plant variety business mirror those described by Milyaeva and Neyland (this volume) for the case of British higher education reform: the premises on which the original framework was built, such as the definition of goods and social roles, had shifted. Still, it is difficult to grasp why seed reuse suddenly became so big a problem for the breeding industry. As one breeder puts it, "Well, farmers have always saved and resown. This farmers' exemption, it has always existed. Okay, I don't know if it's always been—I would say it has always had the same extents. Maybe it intensified a bit when [the former GDR] joined and those giant [farms] there resowed" (interview with wheat breeder, April 2015).

Seed reuse has always been a common practice, especially among smallholders who, after a bad harvest, did not have enough resources to purchase next year's seeds on the market. But the amendment was specifically aimed at big agricultural companies that relied on economies of scale and commercial varieties' ability of stable reproduction to save money. The scope of resowing is difficult to assess in that seed reuse fluctuates annually and farmers' activities are a private business. For the season of 2016–17, an estimated 44 percent of all wheat seed in Germany was farm-saved (Pallauf 2018, 1), which is much less than in Canada and the US with 70 percent, Spain with 85 percent, or Australia with 80–95 percent (Sanderson 2017, 232–233). Still, German wheat breeders are missing out on about half of their potential sales. Breeders, who discount their varieties because they expect revenues over several years in a row, feel that they are left out and demand compensation, while farmers regard seed reuse as one of their traditional privileges. The latter's attitude is aptly summarized in the following declaration by a lobby group:

> Sowing and harvesting, keeping a part of the harvest for resowing it next year—this ancient central principle of agriculture shall now no be longer possible without restrictions. Plant breeders do not only demand—legitimate—license fees when they sell new seeds to the farmers, but they also demand money for the following 25 to 30 years if the farmer is reusing a part of his harvest as seeds—so-called resowing. (IG Nachbau 2016, translated)

To resowing farmers, the nature of seeds is clear: they are an asset. Like a tractor, a harvester, or a plow, they are used for running the farm and not (like fertilizers) completely converted into a more valuable output. But the quote also frames the relationship between farmers and breeders in a particular fashion: while the latter have a legitimate claim to a single monetary compensation for their seeds, farmers consider themselves to be quits (Callon and Latour 1997) after this transaction. In their eyes, all contracts are fulfilled with the acquisition of the original seeds, and there is no need for a long-term relationship through continued license fees and declaration from the side of the farmers.

Resowing farmers thus understand seeds and plant varieties as an on-farm *asset*, where it becomes their lasting property, but as a *commodity* that is exhausted in the transaction at the moment of purchase on the market. They also regard themselves as producers with and owners of, not consumers and renters of seeds. Framing the issue with Slater (2002) as one of alienation, the problem is that too much is alienated when breeders sell their seeds: not just the possibility of using them for cultivation and harvesting but also the ability to multiply them. Since the knowledge necessary for this kind of "piracy" is inseparably built into the commodity itself, the farmer-consumer cannot be deprived of it, giving them an opportunity to become a farmer-reproducer (and, in the breeders' eyes, a bio-pirate). In turn, too little is returned to the breeders. What is compensated is only a portion of the variety paid for in an accordingly discounted single sales transaction. The variety's R&D costs, however, are meant to be covered over several years through recurring purchases, not within one single sales act. Costs of appropriating a new variety are spread across several thousand bags of seed, every single one bearing the potential to recreate the variety at a much smaller cost. In consequence, not all uses of seed are adequately compensated in the eyes of the breeders: those that turn seed into a means of production rather than consumption, they argue, touch upon their intellectual property. In other words, resowing farmers treat seed like a disentangled commodity on the market while breeders consider it an entangled asset of theirs (Muniesa 2008). On the farm, in contrast, farmers regard seed as their reproducible asset whereas breeders wanted them to subject it to consumption only.

The compromise eventually reached favored the breeders, at least on paper. While the right of reuse remained with the farmers (allowing them to resow without prior permission), breeders were granted the right to be included in the valuation of the seeds—independently of farmers' actual revenues. After years of hard-fought negotiations, the price for resowing a protected wheat variety was set at 50 percent of the original license, symbolically accounting for the costs of cleaning and treating the seeds which the farmers bore themselves. If farmers' understanding of varieties as their on-farm assets was acknowledged by the law, it simultaneously framed seeds as breeders' commercial assets that did not simply change hands on the market, but remained part of a larger system of claims and obligations. And yet, the breeders' victory should soon turn out to be a Pyrrhic one.

The Troubles of Doing Property

That breeders were meant to be included in the valuation processes in the fields also meant that the seed market was no longer enough for commercially interacting with the farmers. After the revision of German PVP laws in 1997, breeders needed an apparatus that allowed them to control sowing practices and to enforce their monetary claims. The seed trust administration (Saatguttreuhandverwaltung, STV), until then a barely known institution supported by the breeders, should monitor the use of seeds throughout Germany and collect resowing fees from farmers, if necessary, also by taking legal action. Even so, breeders encountered resistance early on. While resowing fees in wheat rose to around €8 million in the beginning, they soon dropped down to €4 million and are now back at about €9–10 million (interview with wheat breeder, December 2016). Compliance also differed regionally, with 97 percent of all farmers in the state of Thuringia filling out a resowing declaration compared to 23 percent in North Rhine-Westphalia (interview with industry representatives, September 2017). The system's loophole was the lack of identity of seeds: mature plants were readily distinguishable from each other, but once seeds were sacked and shipped, there was no way of telling which was which and, even worse, whose. Breeders' attempts to make farmers declare in questionnaires what, when, and where they had sown were soon stopped by courts. Only in case of reasonable suspicion could they demand information from a farmer.

It was not only farmers, though, who ignored legal boundaries of property. It was also the plants that did not discriminate between breeders and farmers, producers and consumers, asset and commodity. Provided they had been bred via selfing, they could theoretically work in both regimes, but here they sided with the farmers. While the legal boundaries of property were quite clear, a continuum existed in the material realm. Breeders tried to remedy this betrayal by stepping up their surveillance game. Seed processors and distributors were legally obliged to report suspicious batches of seed and, in case of reasonable suspicion, to take samples. Nonetheless, this did not turn out as a success either. First of all, because a lot of seed processors also sold pesticides and fertilizers to farmers and would rather turn a blind eye on undeclared resowing than to risk losing a customer for something they had no financial stakes in. Second, because many larger agricultural operations had their own in-house processing facilities and did not rely on potentially untrustworthy third parties. Breeders' desperate efforts widened the trench between farmers and themselves. They consolidated the farmers' stance toward resowing fees as illegitimate and allowed them to win public sentiment for their side. Reframing property relations in ways that resembled a "refeudalisation" (Schubert et al. 2011) through financial obligations and supervision was met with disapproval in Germany. In 2005, the STV received the German Big Brother Award for its attempts to survey farmers and collect remuneration from them.

In retrospect, breeders admit that their approach back then was not a successful PR strategy. Today, they are attempting to approach the farmers and their representatives, making the argument of a shared interest in a thriving wheat breeding sector and highlighting the benefits farmers receive from buying new seed every year. Following the lesson of "verify but trust," wheat breeders and their associations are currently trying to bridge the gap that the battle over the legal shape of plant varieties has created. These attempts are, however, only partly successful. Revenues from resowing fees keep fluctuating; according to the breeders' calculations, about a third of all resowing farmers do not pay remuneration in cereals, equaling a missing €5–6 million out of a total of €14–15 million in 2014 (Würtenberger 2014, 119). With few legal means and hardly any information about how much resowing is taking place, breeders remain in a weak position. Although wheat breeders see revenues from resowing fees as essential for the future—the primary market alone is too small to sustain the existence of currently seventeen commercial wheat breeding programs—farmers question the economic precariousness of the seed producers. "On every breeder's yard, there's a new BMW and a new Fendt [tractor]; I doubt they're doing that badly," a farmer puts it in a conversation (interview with part-time farmer, March 2016). Faced with such attitudes, wheat breeders are not placing all their bets on the goodwill of farmers and policy makers. Some of them are envisaging a different strategy, hoping it can turn the tables between breeders and farmers.

Back to Commodities: Hybrid Breeding

Inbreeding through selfing has, for a long time, dominated as a breeding method in most crop species. There is a notable exception to this: as early on as the 1930s, corn was turned into a hybrid crop by US geneticists and breeders who could demonstrate the superiority of crosses between two inbred lines over their parental varieties. Hybrid varieties came with considerably more yield, were more uniform in appearance, and promised a bright future for breeding (Fitzgerald 1990). They had another advantage for breeders, however, namely their lack of transgenerational stability. While first-generation (F_1) plants are more homogenous and have higher yields than their parents, subsequent generations exhibit Mendelian patterns, with recessive alleles being expressed, while yield approaches the parental average. This represented an effective biological technique for preventing farmers from resowing, not only because they would forgo yield but also because the plants would lose their commodity characteristics of stability, uniformity, and calculability. It would become harder and harder to predict when to sow and when to harvest, with some plants still being green and others ripe for harvesting. The more the harvest was replanted, the more varied plants would become, making resowing less and less attractive. As Kloppenburg (2004, 97) puts it, hybrid seed is not biologically but "economically sterile." Breeders, in contrast, could still reproduce hybrid varieties because they were in possession of the original parental lines. By being "consumed" over time, hybrid varieties turned into commodities on the farmers' end while remaining assets for the breeders. Where remuneration represents a move for extending the asset character of a breeder's variety to the farm by collecting revenues after the market, hybrid breeding is a move toward more commodification based on preventing assets to slip from the breeder's hands. Hybrid breeding decouples seed from grain (Kloppenburg 2004, 93), asset from commodity. The parental lines, which are held on to (Birch 2017a), not disclosed or handed to third parties, embody and reproduce the asset, while the F_1 seed that is sold to the farmers and can only be consumed represents the commodity.

Historically, hybrid breeding in corn was successful for several reasons. Higher yields allowed to turn hybrid varieties into a win-win scenario for both breeders and farmers; the gain in yields made farmers swallow their skepticism toward being deprived of the ability to resow; and, finally, hybrid breeding was simple in corn—since female and male flowers are separated and rather big, mechanical sterilization was cheap and simple (Fitzgerald 1990). The situation with many other crop species, especially wheat, is different. Wheat flowers are minuscule and delicate, with male and female flowers packed tightly together in the plant's spikelets, requiring skilled and patient work for removing the male parts. Since it is so time- and money-consuming, sterilization of flowers is only done for initial crosses between a few hundred plants; manually producing seed on an industrial scale would be impossible (Whitford et al. 2013). Although so-called gametocides exist, which allow for a chemical sterilization of wheat plants, they are highly toxic, banned in Germany, and patented by a single firm, making their use unattractive (Becker 2011).

Since the 1980s, however, molecular techniques of hybrid breeding have gained ground (Acquaah 2012; Becker 2011). In rye, rapeseed, and barley, cross-fertilization of lines with specific cytoplasm combinations allows for controlled production of hybrids with only around 10 percent contamination by undesired pollen. While many crop species lack "hybrid vigor," the prospect of excluding farmers from resowing has made breeders go for hybrid varieties, despite their production costs being about twice those of conventional varieties (Nickl et al. 2014, 33). In many smaller crop species, breeding companies had an easier time of collectively shifting to hybrids and of simultaneously lobbying public authorities to loosen the thresholds for purity and stability of hybrid seed, as markets were already strongly consolidated. Although farmers were promised higher yields through hybrid vigor and increased breeding efforts funded by higher seed prices, comparisons with some of the still existing inbred varieties casts doubt on these claims (LfL Bayern 2014; Nickl et al. 2014, 2016).

Nevertheless, hybrid breeding remains attractive for wheat breeders. Many public research institutes consider hybrids the future of wheat breeding, citing the same arguments used by rapeseed and rye breeders a few decades ago. While many small and middle-sized companies seek a cheap and efficient method for producing hybrid varieties, they also fear the uncertainties of switching to a new breeding system. So far, the few hybrid wheat varieties on the market are not outperforming inbred ones, and collectively moving toward more expensive hybrid seed represents a challenge in the competitive wheat market. Still, many actors in the wheat breeding industry are optimistic and in favor of hybrid wheat. One breeder at a public research institute leading a program for the development of a hybrid system especially stressed the openness of such a system, which would be available to all companies without license fees, not only the big firms with their own R&D departments, and thus reconcile property protection with scale neutrality (interview with wheat breeder, April 2015).

That such scale neutrality is difficult to achieve was demonstrated a few months after that breeder made that statement. When asked for their opinions, the private project partners who had come to observe the results and discuss the future of the hybrid program remained silent, until one among their ranks voiced his skepticism. Shortly before, the Swiss biotech company Syngenta had had to compensate farmers after one of its hybrid barley varieties had been found to be contaminated with foreign seed. Barley has only half the acreage of wheat in Germany (1.6 vs. 3.2 million hectares in 2017; Statista 2018). Still, the compensation paid by Syngenta was ten times his own company's total balance sheet, that breeder exclaimed. Even if the joint program were to yield a feasible and cheap method for producing hybrid seed, he could not possibly take the risk of going bankrupt over hybrid varieties. Consequently, the institute's hybrid program will fall dormant. Some basic research will probably still be conducted, but the development of a working technology is off the table for now. Elsewhere, though, public and private actors are still working on making hybrid wheat the future, especially within the plant science multinationals. To make up for its higher production costs and to compete with cheaper inbred seed, however, hybrid wheat varieties would need a stable yield advantage of around 6 percent. Since heterosis is so low in wheat, none of the existing varieties currently manages to come close to this (interview with wheat breeders, August 2018). Some industry experts are therefore pessimistic about the commercial prospects of the technology (interview with former biotech executive, June 2018).

At first glance, it seems like hybrid breeding could successfully combine the advantages of both asset- and commodity-shaped valuation processes. It would turn wheat varieties into assets on the breeders' side, securing them long-term revenues from selling seeds. In contrast, seed would only be consumed (and thereby commodified) on the farmers' side as they would be unable to recreate hybrid varieties. In spite of that, financial costs and risks of this technological solution are usually not discussed publicly. For small to medium-sized breeding companies, they are high enough to stick with traditional inbreeding, which is both cheaper and more secure. Bigger biotech-oriented firms that have only recently entered the German market still hope that, in the long run, hybrids will prevail, lead to higher revenues, and disentangle property relations between breeders and farmers. Currently, however, hybrid wheat is not economically convincing for any of the involved parties.

To Commodify or to Assetize?

The case of wheat seeds shows that telling assets from commodities is not that easy. That lawmakers think of plant varieties as a form of intellectual property and that breeders calculate their expenses against varieties as a lasting source of revenues commercialized at discounted rates does not mean that they do not also appear as commodities at other times or even simultaneously. Quite to the contrary: their valuation through markets requires a certain degree of commodification, both material and semiotic, channeling their value and allowing for an easy conversion to money. Varieties cannot do without certified, standardized bags of certified, standardized seed. There are, of course, also assets which are subject to less calculation and more speculation. Nevertheless, I would argue that the overwhelming majority of assets only work when coupled with some form of market where they can quickly and with little risk be converted into money through the exchange of commodities for money. In this sense, and contrary to Birch and Muniesa's (important) provocation in the introduction to this volume, commodification is not dead. It might, however, have taken on a different life.

The commodification of assets does, in turn, not imply that every aspect of them is subject to calculation and alienability. For example, although the costs of developing a new plant variety are usually estimated at around €/$1–2 million (Goodman 2002, 30), the specific costs for one particular variety cannot easily be assessed, simply because they escape straightforward zero-sum games of production and consumption:

> At the moment it is said that breeding a new variety costs 1.5 million, 1 to 1.5 million. And that's also relative. If I say, okay, we invest 700,000 per year or 800,000 here…yeah, maybe just 700,000 for wheat only. Barley, you'd have to subtract all of that, subtract [our barley breeder]. We invest 700,000 for the wheat program [here]. And you got three varieties admitted in a year, or maybe just one, then getting one variety admitted costs you between 300,000 and 700,000 euro. So that's really difficult to say. (interview with wheat breeder, April 2015)

Seed price is not simply the sum of cost factors. For the breeder quoted above, who licenses his seed at the upper end of the price range, it also carries a signaling function. Convinced that his varieties are of a higher quality and require less farming input than most others, he holds that a cheap product is a worthless product.

To "have a good variety" that will last a long time, enjoy popularity among wheat farmers, and yield considerable income which will not simply go into covering past expenses, but also future projects is important to breeders. But it is not only the breeders who decide whether seeds will be a commodity or an asset. Farmers, infrastructures, legislation, and not least the wheat plants themselves have a considerable say in that matter. Combining the advantages of assets—lasting rents instead of recurring labor efforts—with the convenience of commodities—easy, straightforward and almost automated commercialization—bears risk. Breeders cannot guarantee that their product will behave like the asset they want it to be once it changes hands. In wheat breeding, it is the mutability of biological beings which undermines the articulation of assetization and commodification. If varieties at the farmer's end of the valuation chain lose their commodity character, assets become a problem for the producers. While we take it for granted that commodities are consumed not only in consumption, but that they actually change hands as a whole on the market, customer complaints do not only make authorship and claims but also responsibility reemerge.

Hybrid breeding demonstrates why—somewhat contra Birch's (2017a) prompt—it is still important to look at the material dimension of goods. Theoretically, there are ways of separating asset and commodity, of dividing the world into producers who hold the former and consumers who have to acquire the latter again and again. In practice, the failure of wheat breeders to achieve this separation can be located in the fickle nature of wheat genomes, the delicate form of cereal flowers, and the very calculable limits of a medium-sized firm. On the other side of the spectrum, hybrid breeding in corn and other crop species has become an effective tool for extracting higher rents from farmers who pay a premium for seed they can no longer appropriate as an asset. This urges us to think about the specific articulation of goods through material, social, legal, and other techniques. Property plays an important (and, as I believe, so far largely neglected) role in this context: Which part of a good should actually be sold? What should be retained? What can be alienated, what can be appropriated by the other? How can legal, material, and social aspects be made to agree with each other, and what happens if they disagree?

That assets and commodities may be tightly interlinked along a value chain should also not distract us from noticing that turning varieties into assets is not an isolated event or idea. New forms of property are entering the plant breeding sector, disrupting established social and economic relations. The 2010 Nagoya Protocol has made nation states and indigenous communities owners of their domestic biodiversity, previously subject to the public domain (Brand and Vadrot 2013). Biodiversity is now something to be managed—not just for conservation but also for economic reasons. For states like Mexico, Ecuador, or Namibia, it has become a national asset promising a bright economic future in the age of bioprospecting (Hayden 2003; Heeren 2016, 2017). Downstream the value chain, the protocol's implementation in Germany and other EU countries has led to concerns among breeders. The obligation to pay royalties for the use of foreign germplasm implies that such material as well as all other sources of genetic diversity used for breeding have to be minutely documented. If they do not comply, some national laws include heavy sanctions for breeders, from five-digit fees to imprisonment. Intended to foster exchange of genetic material and creating more legal certainty by turning states into owners of their biodiversity, Nagoya currently has the opposite effect.

Meanwhile, Europe has also seen a surge in plant patents granted by the European Patent Office (EPO) lately, especially in vegetables (Parthasarathy 2017). While patents are nothing new for the US seed sector, they have deliberately been kept out of European plant breeding by legislation until recently. In Europe, plant varieties are not eligible for patent protection, only for PVP. For both breeders and farmers, this is important since PVP is less restrictive than patent protection, giving them more freedoms in using foreign seed as a parent. Patents, in contrast, offer valuation channels very different from commodification, as Hyo Yoon Kang and Victor Roy show in their contributions to this volume. Critics argue they have become an outlet for turning science and technology into money without having to go through the troubles of production and consumption at all. At the same time, patents are often too complex and laden with transaction costs to be easily commodified. This is one of the factors suspected behind the rapid concentration of the US seed sector following the introduction of patents on genetically modified plants (Schenkelaars et al. 2011). In the same case, it has also led to the replacement of the commodity form—that is, disentangled goods and market transactions—with lasting contractual obligations between biotech firms and farmers, supplemented by an extensive socio-technical surveillance infrastructure (Schubert et al. 2011; Müller 2015).

Nevertheless, there are, also attempts at recommodifying plant patents. The Swiss biotech giant Syngenta, for example, introduced the platform *Traitability* in 2012, which offers access to the company's patented vegetable traits on simple, predefined, and purely monetary terms. Two years later, a cross-industry initiative spearheaded by Syngenta launched a patent clearinghouse for non-biotech patents on vegetables meant to reduce transaction costs and uncertainty around patents (Bjørnstad 2016). It is therefore still unclear what business vision companies connect to plant patents. After protests and an explicit disapproval by the EU commission, the EPO has backed down from its liberal position in granting patents (EPO 2017), a decision overruled shortly after by its own Enlarged Board of Appeals. The property question in European plant breeding is thus still open.

Conclusion

The examples discussed in this chapter point to a trend toward what can be aptly termed "assetization" of property, not just in plant breeding but also in other fields (Perzanowski and Schultz 2018). In the market economy, private property, with its extensive liberties, perfect alienation, and lack of lasting obligations, has traditionally been a necessary prerequisite for the exchange of goods as commodities. Unlike landed property and other forms of ownership centered around managing possessions and extracting value from them (see Nadaï and Cointe, this volume), the logic of private property on the market is fixated on alienation (Slater 2002), letting go of what you own and being adequately compensated in return (Appadurai 1986). Aside from the disruption this idea has historically caused (Polanyi 2001), one of its effects was a great liberty in property, allowing people to leave behind the constraints of their social context and dealing freely with strangers.

Understanding property as an asset, however, implies treating it not as something to be alienated (and, consequently, reappropriated), but something that can be let go of and yet returns to its owner. The specific articulation with commodities is crucial: it needs to provide all the advantages of commodity exchange—calculability, easy transport, scalability, straightforward appropriation—without what would be a disadvantage from an assets perspective—complete alienation and being quits. There are many ways of making goods, people and money return to the asset holder—from a legal-technological surveillance and sanction infrastructure to a more concentrated product like hybrid seed which will obey the seller but not the buyer. This is why Birch's (2013) call to bring together STS and political economy is so important: we will only understand the differences between specific articulations of assets and commodities by looking closely at the political-economic effects a certain arrangement has. Through increasing yield and improving seed performance, hybridization can assetize varieties in corn and protect them from infringement. It is not suited, however, for protecting transgenes in the same varieties (Pottage 2011, 109f.)—the latter require an extensive apparatus of patents, snitch lines, technology use agreements and biomolecular detection techniques in order to work as assets (Schubert et al. 2011), making them socially and economically much more expensive. In the German wheat market, assetization of seed through a mixture of cooperation, dialogue, surveillance, and PR has been a moderate success in the last years, with currently some 70 percent of all seed-saving farmers paying remuneration (interview with wheat breeder November 2018; interview with industry representatives September 2017). The missing 30 percent of resowers point to the limits of assetization as a way of framing both economic calculations and economic exchange and of the control breeders can exert over "their" assets.

Assetization is neither a given nor something that is simply conceived, planned, and executed by economists, investors, managers, or multinationals. As a reaction to wheat varieties experiencing changes in their asset value and character through shifts in the economic landscape of farming and breeding, breeders attempted to reassetize their seed legally, biologically, and economically. But the success of such attempts depends on the compliance of consumers, laws, markets, competitors, genomes, and many other factors which all too often escape calculation or Latourian enrollment. Like commodification, assetization remains performative, leaving spaces and opportunities for intervention. A lot of things can be turned into assets and commodities, but commodities and assets can still be turned into other things—with or without the consent of their producers.

Notes

1. This chapter is based on participant observation, document analysis, and interviews with stakeholders in Germany, the Netherlands, and Switzerland between 2015 and 2018.

2. In 2017–2018, a rough estimate for the price composition of a quintal (100 kg) of certified winter wheat seed would be €20 for the base price or "consumption" (i.e., value as raw material for processing or fodder), €6.25 for the seed multiplier's share, €10 for processing and preparing (e.g., coating) of seed, €2 for the bag, €1,75 for distribution and sales, and €7–13.25 for the breeder's license (interview with wheat breeder, November 2018). Somewhat more precise but also more dated figures for a €39.80 bag of 2009/10 cereal seed (DLZ 2010) are €10.25 base price (26 percent), €10.10 for coating and bagging (25 percent), €6.30 for the multiplier's share (16 percent), €5.40 for sales and distribution (14 percent), and €7.30 for the breeder's license (19 percent).

References

Acquaah, G. 2012. *Principles of Plant Genetics and Breeding.* 2nd ed. Hoboken, NJ: Wiley.

Aistara, G. 2014. Latvia's Tomato Rebellion: Nested Environmental Justice and Returning Eco-Sociality in the Post-Socialist EU Countryside. *Journal of Baltic Studies* 45 (1): 105–130.

Appadurai, A. 1986. Introduction: Commodities and the Politics of Value. In *The Social Life of Things: Commodities in Cultural Perspective,* edited by A. Appadurai, 3–63. Cambridge: Cambridge University Press.

Becker, H. 2011. *Pflanzenzüchtung.* 2nd ed. Stuttgart: Ulmer.

Birch, K. 2013. The Political Economy of Technoscience: An Emerging Research Agenda. *Spontaneous Generations: A Journal for the History and Philosophy of Science* 7 (1): 49–61.

Birch, K. 2017a. Rethinking Value in the Bio-economy: Finance, Assetization, and the Management of Value. *Science, Technology, & Human Values* 42 (3): 460–490.

Birch, K. 2017b. Towards a Theory of Rentiership. *Dialogues in Human Geography* 7 (1): 109–111.

Bjørnstad, Å. 2016. "Do Not Privatize the Giant's Shoulders": Rethinking Patents in Plant Breeding. *Trends in Biotechnology* 34 (8): 609–617.

Brand, U., and Vadrot, A. 2013. Epistemic Selectivities and the Valorisation of Nature: The Cases of the Nagoya Protocol and the Intergovernmental Science-Policy Platform for Biodiversity and Ecosystem Services (IPBES). *LEAD: Law, Environment and Development Journal* 9 (2): 202–220.

Brandl, B. 2017. *Wissenschaft, Technologieentwicklung und die Spielarten des Kapitalismus: Analyse der Entwicklung von Saatgut in USA und Deutschland.* Wiesbaden: Springer VS.

Brandl, B., and Glenna, L. 2017. Intellectual Property and Agricultural Science and Innovation in Germany and the United States. *Science, Technology, & Human Values* 42 (4): 622–656.

Callon, M. 1998. Introduction: The Embeddedness of Economic Markets in Economics. *Sociological Review* 46 (S1): 1–57.

Callon, M. 1999. Actor-Network Theory—The Market Test. *Sociological Review* 47 (S1): 181–195.

Callon, M., and Latour, B. 2011. "Thou Shall Not Calculate!" or How to Symmetricalize Gift and Capital, http://bruno-latour.fr/sites/default/files/downloads/P71_CAPITALISME-MAUSS-GB.pdf.

Callon, M., and Muniesa, F. 2005. Peripheral Vision: Economic Markets as Calculative Collective Devices. *Organization Studies* 26 (8): 1229–1250.

Demeulenaere, E. 2014. A Political Ontology of Seeds: The Transformative Frictions of a Farmers' Movement in Europe. *Focaal* 2014 (69): 45–61.

Demsetz, H. 1967. Toward a Theory of Property Rights. *American Economic Review* 57 (2): 347–359.

DLZ. 2010. Viele verdienen mit. *Dlz Praxis* 10: 36–39.

EPO. 2017. EPO Clarifies Practice in the Area of Plant and Animal Patents. June 29. https://www.epo.org/news-issues/news/2017/20170629.html.

Fitzgerald, D. K. 1990. *The Business of Breeding. Hybrid Corn in Illinois, 1890–1940.* Ithaca, NY: Cornell University Press.

Goodman, M. 2002. New Sources of Germplasm: Lines, Transgenes, and Breeders. Presented at the Memoria Nacional de Fitogenetica, Saltillo, Coah.

Graeber, D. 2011. Consumption. *Current Anthropology* 52 (4): 489–511.

Hardie, I., and MacKenzie, D. 2007. Assembling an Economic Actor: The *Agencement* of a Hedge Fund. *Sociological Review* 55 (1): 57–80.

Harwood, J. 2012. *Europe's Green Revolution and Others Since. The Rise and Fall of Peasant-Friendly Plant Breeding.* London: Routledge.

Hayden, C. 2003. *When Nature Goes Public. The Making and Unmaking of Bioprospecting in Mexico.* Princeton, NJ: Princeton University Press.

Head, L., Atchison, J., and Gates, A. 2012. *Ingrained: A Human Bio-geography of Wheat.* Farnham, UK: Routledge.

Heeren, A. 2016. Commercialization of Biodiversity: The Regulation of Bioprospecting in Ecuador. *fiar* 9 (2): 94–117.

Heeren, A. 2017. Globale Bioökonomien. Gesellschaftliche Naturverhältnisse im Kontext der Bioprospektion in Namibia. In *Geschlossene Gesellschaften. Verhandlungen des 38. Kongresses der Deutschen Gesellschaft für Soziologie in Bamberg 2016*, edited by S. Lessenich. Essen: Deutsche Gesellschaft für Soziologie.

IG Nachbau. 2016. Willkommen bei der IG Nachbau, http://www.ig-nachbau.de/start/.

Kang, H. Y. 2015. Patents as Credit. When Intellectual Property Becomes Speculative. *Radical Philosophy* 194: 29–37.

Kempf, H. 2016. Weizenzüchtung bei der Secobra Saatzucht im Spannungsfeld zwischen Wettbewerb, Landwirtschaft, Handel und Verbraucher. Presentation Given at the Department für Nutzpflanzenwissenschaften, Universität Göttingen.

Kloppenburg, J. 2004. *First the Seed: The Political Economy of Plant Biotechnology*. Madison: University of Wisconsin Press.

Kopytoff, I. 1986. The Cultural Biography of Things: Commoditization as Process. In *The Social Life of Things: Commodities in Cultural Perspective*, edited by A. Appadurai, 64–94. Cambridge: Cambridge University Press.

Latour, B. 1987. *Science in Action: How to Follow Scientists and Engineers through Society*. Cambridge, MA: Harvard University Press.

Latour, B. 1993. *The Pasteurization of France*. Cambridge, MA: Harvard University Press.

LfL Bayern. 2014. Vergleich der Ertragsleistung von Linien- und Hybridsorten bei mehrzeiliger Wintergerste, https://www.lfl.bayern.de/mam/cms07/ipz/dateien/13_hybr_linie.pdf.

Müller, B. 2015. Fools Gold on the Prairies: Ontologies, Farmers and Their Seeds. *Tsanta* 20: 61–73.

Muniesa, F. 2008. Attachment and Detachment in the Economy. In *Attachment: Sociology and Social Worlds*, edited by P. Redman, 111–142. Manchester, UK: Manchester University Press.

Muniesa, F., Doganova, L., Ortiz, H., Pina-Stranger, À., Paterson, F., Bourgoin, A., Ehrenstein, V., Juven, P.-A., Pontille, D., Saraç-Lesavre, B., and Yon, G. 2017. *Capitalization: A Cultural Guide*. Paris: Mines ParisTech.

Nickl, U., Hartl, L., and Herz. M. 2014. Erträge steigern mit Hybridsorten? In *Ackerbau—mit hohen Erträgen erfolgreich wirtschaften*, edited by LfL Bayern, 33–40. Freising-Weihenstephan: LfL Bayern.

Pallauf, M. 2018. *Planning and Decision-Making by Seed Trading Companies for Winter Wheat (Triticum aestivum) Multiplication: A Mixed-Methods Analysis for Bavaria*. Weihenstephan: TU Munich.

Parthasarathy, S. 2017. *Patent Politics: Life Forms, Markets, and the Public Interest in the United States and Europe*. Chicago: University of Chicago Press.

Perzanowski, A., and Schultz, J. 2018. *The End of Ownership: Personal Property in the Digital Economy*. Cambridge, MA: MIT Press.

Polanyi, K. 2001. *The Great Transformation. The Political and Economic Origins of Our Time*. 2nd Beacon Paperback ed. Boston: Beacon.

Pottage, A. 2011. Biotechnology as Environmental Regulation. In *Law and Ecology: New Environmental Foundations*, edited by A. Philippopoulos-Mihalopoulos, 105–125. Oxford: Routledge.

Sanderson, J. 2017. *Plants, People and Practices. The Nature and History of the UPOV Convention*. Cambridge, MA: Cambridge University Press.

Schenkelaars, P., de Vriend, H., and Kalatzaidonakes, N. 2011. *Drivers of Consolidation in the Seed Industry and Its Consequences for Innovation*. Bilthoven, NL: COGEM.

Schubert, J., Böschen, S., and Gill, B. 2011. Having or Doing Intellectual Property Rights? Transgenic Seed on the Edge between Refeudalisation and Napsterisation. *European Journal of Sociology* 52 (1): 1–17.

Slater, D. 2002. From Calculation to Alienation: Disentangling Economic Abstractions. *Economy and Society* 31 (2): 234–249.

Statista. 2018. Anbaufläche von Getreide in Deutschland nach Art bis 2017, https://de-1statista-1com-10013474w04e4.emedia1.bsb-muenchen.de/statistik/daten/studie/28887/umfrage/anbauflaeche-von-getreide-in-deutschland-seit-1960/.

STV. 2018. Vertragssortenliste 2017/18, https://www.stv-bonn.de/sites/default/files/downloads/2018-02/NBE18_Vertragssortenliste_final.pdf.

Whitford, R., Fleury, D., Reif, J., Garcia, M., Okada, T., Korzun, V., and Langridge, P. 2013. Hybrid Breeding in Wheat: Technologies to Improve Hybrid Wheat Seed Production. *Journal of Experimental Botany* 64 (18): 5411–5428.

Würtenberger, G. 2014. Nachbauvergütungen: Eine kritische Bestandsaufnahme. In *Rechtsschutz von Pflanzenzüchtungen. Eine kritische Bestandsaufnahme des Sorten-, Patent- und Saatgutrechts*, edited by A. Metzger, 105–114. Tübingen, DE: Mohr Siebeck.

9 Turning Nature into an Asset: Corporate Strategies for Rent-Seeking

Les Levidow

Introduction

Since the turn of the twenty-first century, financial concepts have become prevalent in managing natural resources. For example, UN institutions have increasingly promoted market-type instruments such as carbon credits, biodiversity offsets, payment for ecosystem services (PES), and so on. Such instruments involve financial transactions meant to reduce or compensate for environmental harm.

Beyond financial transactions, the metaphor of "natural capital asset" has been used to highlight the source of ecosystem services on which all economic activities depend. Together they face threats of resource degradation, resulting from the economic "invisibility of nature" (UNEP 2011, 16; TEEB 2008). To render these assets more visible, natural capital accounting (NCA) has been elaborated for three different contexts: nature conservation policy, tradable permits, and business strategies. In the latter context, a multi-stakeholder coalition has been devising methods for companies to assess their dependence on natural capital assets as a basis for future corporate strategies (e.g., NCC, 2016, 2018). This chapter investigates this process and asks: How does corporate NCA turn nature into an asset? With what drivers, roles, and stakes?

The chapter has the following structure: it first outlines theories of assetization and their valuation, especially nature; then agendas for reifying nature as an asset and the "global justice" critiques of this agenda; then a business protocol to operationalize natural capital accounting (NCA), with specific companies that have done so; then business strategies for stewarding natural capital as "shared assets." It concludes by answering the above questions.

Nature as Asset: Theoretical Perspectives and Research Methods

For several centuries, nature has been characterized by various economic metaphors, each reifying human constructs and processes as eternal things in various ways. In late eighteenth-century Germany, for example, the complex biological dynamics of forest stands, the basis of natural forest regeneration, were simplified for a more efficient "rational forestry." This scientific-managerial model served to maximize wood production for greater economic returns through tree plantations, thus redefining forests (Scott 1998, 11–22).

Nature has had shifting personifications since the eighteenth century. Until then it was characterized as an organism, an anthropomorphic projection of communities maintaining commons, whereby everyone's fair access was understood as a natural right. With the enclosure of commons, however, nature was recast through metaphors of a mechanism and market (Williams 1980); land was turned into an asset for capital accumulation. Across history nature metaphors "have ranged from inherent and inevitable bitter competition to inherent mutuality or co-operation," each version eternalizing human behavior (Williams 1983).

In the nineteenth century, evolutionary theory understood ecological niches and species change through market-like competition as an anthropomorphic projection: "All organic beings are striving to seize on each place in the economy of nature," argued Darwin (1859, 90). His theory naturalized capitalist markets, as Karl Marx observed: "Darwin did not know *what a bitter satire* he wrote *on mankind*, and especially on his countrymen, when he showed that free competition—the struggle for existence, which the economists celebrate as the highest historical achievement—is the normal state of the animal kingdom" (Marx 1862, 156–57, emphasis in original). Those successive metaphors have served private-interest claims on natural resources. This literature survey focuses on the asset-stocks metaphor of nature, toward analyzing corporate NCA in subsequent sections.

Assetization as Rent-Seeking

As in some examples above, ontological metaphors reify a process as a thing. For example, viewing price inflation as an entity allows us "to quantify it, to identify a particular aspect of it, see it as a cause, act with respect to it, and perhaps even believe that we understand it" (Lakoff and Johnson

1980, 26–27). Through everyday experience, some ontological metaphors become taken for granted, seen as natural rather than as metaphors.

Such naturalization is illustrated by carbon credits. As a financial metaphor, credit originally denoted a social relationship—confidence that a creditor would repay a debt. Credit was later reified as a quantifiable, interchangeable thing through the carbon-credit scheme. To implement the 1992 UN Framework Convention on Climate Change (UNFCCC), the 1997 Kyoto Protocol created tradable carbon credits, entitling permit holders to pollute the atmosphere. The institutional shift divided NGOs. Some helped develop the financial infrastructure of carbon markets (Andonova and Hoffmann 2012, 59), but their involvement provoked criticism from other NGOs and especially social movements in the global South. A contingent reason for this criticism was the fall in the carbon price, which weakened the incentive to reduce greenhouse gas (GHG) emissions.

More fundamentally, some critics denounced the scheme for "commoditizing nature" (see later section). In practice, commodity production generally adds to the surplus value which can be realized through market exchange. By contrast, carbon credits entitle the owner to emit GHGs, as necessary conditions of production. Their purchase thereby imposes an extra cost, redistributing value which already exists.

Hence, carbon credits are analogous to a "form of rent" rather than new commodities: "The distribution and circulation of these entitlements through market-based mechanisms should not lead us to treat them as 'commodities,' but rather as a form of rent.... Emissions rights do have an exchange value and a use value, but they do not represent value" (Felli 2014, 254, 268). This analysis draws on Marx's concept of rent—"the price paid to the owner of natural forces or mere products of nature" for the right to use them (Felli 2014). As necessary conditions of production, carbon credits relate to the right to use the environment as a pollution sink.

A cap on GHG emissions would restrict a key condition of production, so it became a focus of rival capitalist claims. For the 1992 UNFCCC, restricted access to the sink was foreseen as incentivizing capital flight: "Governments across the world have faced the contradiction between the need to ensure the reproduction of the conditions of production (and of social reproduction), which would mandate drastic reductions in GHG emissions, and the need to ensure that they retain capital within their boundaries, which generally requires as few regulations as possible" (Felli 2014, 258).

Although the protocol sets a political limit on global emissions, their distribution among producers is effectively "regulated by the law of value"—that is, financial power—thus depoliticizing the global use of space and resources. This rent-seeking role complements the neoliberal project of allocating resources according to polluters' ability to pay rather than regulatory criteria (Felli 2014, 274). This role complements the wider neoliberal agenda of "non-decision making in economic processes"—in other words, the state relegating policy decisions to market exchanges among formally equal parties (Felli 2014, 655).

The neoliberal agenda more generally has promoted rentiership—rent-seeking through various forms of proprietary control. As noted by David Ricardo, rent denotes the transfer of profit from capitalist to property owner, rather than being the basis for generating more overall profit. Today rent more generally denotes ownership and control over a resource, such as nature, knowledge, financial assets, or technology (Birch 2019).

For example, R&D expenditure has been recently understood "as creating an asset with annual depreciation costs, i.e., as capitalized property, meaning that the value of R&D spending will stretch beyond its immediate contribution to production" (Birch 2017, 169). Control over the priorities and commercial use of such knowledge becomes a proprietary matter: "Knowledge assets give owners exclusion rights *and* use rights for copies derived from the asset" (Birch 2017, 172, emphasis in original). While an asset status could justify greater state investment for the public good (see Milyaeva and Neyland, this volume), it has instead been conflated with private interests, which devise rent-seeking strategies to restrict or exclude rivals' access.

Such rent-seeking has several financial forms, including capital ownership, tradable credits, offsets, and payment for ecosystem services. More broadly, rent gains a performative basis in nature-as-asset.

Nature-as-Asset Warranting Investment

Mainstream agendas for sustainable development and natural-resource protection have undergone a shift over the past two decades. The 1990s saw debates over conflicts between economic growth versus social equity and ecological resilience, but more recently such conflicts have been obscured. Instead, economic growth has been relegitimized as a crucial basis for the fair distribution and protection of natural resources. From the earlier focus on state planning and regulation, responsibility has been shifted to municipalities, private companies, and NGOs. Moreover, environmental protection has undergone a commodification process through market-type values, language, and instruments (Gómez-Baggethun and Naredo 2015).

This process has generated new conflicts. For example, financial instruments have expanded the scope for "green-grabbing" in the global South. Here, land grabs have been justified as resource protection, for instance through carbon credits, biodiversity credits, biofuels quotas, payment for ecosystem services, or related offsets (Fairhead et al. 2012). Such instruments reduce the environment to standard measures of resource accounting; this obscures the resource needs and uses of nearby communities, thus serving to dispossess them (Forsyth and Sikor 2013).

Through tradable permits, multiple environmental values are reduced to a single unit of valuation, thus homogenizing those values. This has been theorized as a broader strategy of "depoliticization by economization" (Adaman and Madra 2014). Yet the opposite can happen. As is well documented, REDD+ forest credits have often intensified political conflicts over natural resources and land use, thus jeopardizing corporate reputations. Nevertheless, the scheme expands because they offer various benefits to companies buying the carbon credits.

As motives for carbon credits, companies seek to support environmental claims for corporate supply chains and thus for corporate social responsibility (CSR), or they seek to offset risk from investments in potentially stranded carbon-intensive assets (Laing et al. 2015, 3–4). In that regard, "green supply chain" has become a new buzzphrase replacing or supplementing "sustainable production and consumption." Companies recognize the need to protect their brands and sales through environmental claims such as carbon neutrality and deforestation-free (Kill 2016, 114).

Analogous instruments have been devised for nature conservation over several decades. The 1992 Convention on Biological Diversity (CBD) was foreseen as potentially limiting businesses' access to natural resources, especially through regulatory constraints and benefit-sharing arrangements. To avoid such limitations, business has cooperated with non-governmental conservation organizations (NGCOs) to develop entrepreneurial strategies around market mechanisms for allocating access to sites of nature as capital, such as schemes paying for ecosystem services (Robertson 2006). Funds from US AID and the UN's Global Environmental Facility helped expand the roles of these NGCOs, especially the International Union for Conservation of Nature (IUCN), World Wildlife Fund (WWF), and Conservation International (MacDonald 2010).

Companies seek such partnerships "as necessary risk management to protect their reputations and markets and as a way to open up new markets" (Robinson 2012, 969). Through joint biodiversity initiatives, NGCO-business partnerships have helped to give companies the imprimatur of environmental stewards. Meanwhile NGCOs themselves have become corporate-like entities (Corson 2010).

When such nature-conservation partnerships elaborated the natural capital metaphor, this was critically analyzed as equating nature with financial investment and its potential returns. Indeed, nature becomes personified as a billable service provider (Sullivan 2013; cf. Williams 1980). Valuing natural capital "makes nature legible by abstracting it from social and ecological contexts and making it subject to, and productive of, new market contexts" (MacDonald and Corson 2012, 159). Capital metaphors, moreover, imply that "the environment can be considered, valued, and managed as an asset like any other" (Coffey, 2015, 215). According to Akerman (2005, 48–49), "Instead of approaches which would give a new insight into the evolving everyday practices through which humans are connected with their natural environment, the concept of natural capital seemed to marginalise these discourses and strengthen the ahistorical and non-contextual view of environmental problems."

By highlighting and naturalizing a market return on investment, the nature-as-asset metaphor favors specific types of economic activity and stakeholder participation. Neoliberal conservation practices have a chameleon-like flexibility in creating "both environmentally and market-friendly subjects" (Holmes and Cavanagh 2016, 204). Local people "are increasingly now being incorporated into conservation every time they conduct their new conservation-friendly livelihood activities." Many initiatives have been "working within the lives of rural people, changing their behaviour"; appeals to economic rationales, especially for a return on investment, end up displacing other social values and bonds (Holmes and Cavanagh 2016, 206).

Those critiques extend earlier critical perspectives on neoliberal conservation and more generally neoliberalizing nature. The latter "involves the privatization and marketization of ever more aspects of biophysical reality, with the state and civil society groups facilitating this and/or regulating only its worst consequences"; such ecological fixes are devised in the name of remaking, conserving, or expanding nature (Castree 2008, 142–43, 150). Internal state contradictions are addressed "by off-loading responsibilities to the private sector and/or civil society groups," as if they could provide environmental protection in lieu of interventionist states (Castree 2008, 146, 149).

As critical academics argue, ecosystem degradation has worsened because extractive industrial development models benefit some private interests at the expense of the public good, which lacks adequate protection (see Gilbert this volume). Evading those systemic drivers, natural capital agendas instead attribute ecosystem degradation to a cognitive deficiency—natural capital being financially invisible—whose remedy lies in a holistic valuation and management. This agenda extends broader historical patterns: capitalist environmentalisms "generate their own imaginary featuring powerful managerial agents, situated above both nature and society, that can step in to govern their mutual relations." In this imagined society/nature binary, non-natural humans produce commodities from a supposedly nonhuman nature (Lohmann 2019, 235–236).

As a metaphor, capital implies investment in something that could yield a return. "In financial parlance, capitalisation is about envisaging the value of something in the terms of an investment" (Muniesa et al. 2017, 11–12). This concept gains commonsense force from its metaphorical meaning—getting the most out of something or maximizing advantage—thus thinking like an investor. In order to capitalize on something, "It must be either considered an asset, or turned into one" (Muniesa et al. 2017). Financial valuation plays a performative role in defining something as an asset. Valuation concerns "a relational property of objects," which is performed in the process (Dewey 1939, 5). Extending that insight, capital valuation has a performative role conferring value on something (Muniesa 2012, 33), thus warranting investment. This role has been deeply naturalized, taken for granted, while shaping specific forms of capital (Muniesa et al. 2017, 13).

Research Methods

Drawing on those theoretical perspectives, this chapter analyses how corporate natural capital accounting (NCA) turns nature into an asset warranting financial investment and private-sector stewardship. This study began by analyzing various metaphorical and practical meanings of natural capital, especially changes in meaning since UN bodies gave the concept a higher profile around 2010. Sources analyzed included stakeholder and policy documents, such as UNEP, TEEB, NGCOs, NGOs, business organizations, and consultancies advising them. Preliminary analysis provided a basis for interview questions with nine individuals, mainly around natural capital alliances, during 2015–2016.

Various agendas for NCA have generated significant debate where both sides imply that NCA has direct effects: proponents emphasize the ecosystem benefits, while opponents emphasize harms. But its role cannot be understood as directly causal, because NCA remains embryonic and because it functions within a wider strategic process. Hence, interview questions asked how actors use or understand NCA's key terms (e.g., capital, asset, shared, natural resources, ecosystem services, supply-chain changes, and regulatory standards). They also asked about environmental stewardship as multi-stakeholder engagement. In the light of interviewees' answers, the study reexamined the above sources. This method helped to identify multi-stakeholder processes and their internal tensions, beyond public disputes over NCA (see table 9.1).

Reifying Nature as an Asset: From TEEB to Business Strategies

As a flexible metaphor, natural capital has been given diverse meanings and roles over several decades (e.g., Pearce et al. 1989; MEA 2005; Porritt 2006). It was originally promoted as a persuasive tool for environmental protection. According to the leading UK environmentalist Jonathan Porritt, for example, "If there is any genuinely sustainable variant of capitalism, then it will need to work within the conceptual and linguistics conventions that people are now so familiar with," such as extending financial to natural capital (Porritt 2006, 113).

For development theorists, natural capital has meant a resource empowering community development for better livelihoods through sustainable development. Conversely, human capital enhances natural capital within a wider framework for understanding the institutional design and societal governance of natural resources. A multifaceted capital became a ready metaphor to capture the range of enabling conditions that development conservation advocates might promote in order to achieve desired aims (Wilshusen 2014, 129). Such metaphors highlighted various forms of labor and power relations, but these became obscured by later versions of natural capital.

Initiated by the G7, The Economics of Ecosystems and Biodiversity (TEEB) studies have promoted natural capital evaluation, understanding natural resources as assets delivering flows of ecosystem services, analogous to financial capital yielding dividends. Here, ecosystem services flow from natural capital, seen as stocks or assets—separate from human activities, except for maintenance and restoration costs (TEEB 2008, 32, diagram). Given that "you cannot manage what you do not measure," these studies argue that governments must promote "ecosystem-biodiversity accounting in physical and monetary terms" (TEEB 2008, 6, 54).

As a rationale for such accounting, nature is portrayed as the "GDP of the poor"; poverty is worsened by ecosystem loss. So this must be alleviated in order to ensure "the right of the world's poor to livelihood flows from nature which comprise half of their welfare or more" (TEEB 2008, 4–5, 31). As a plea for fairness, "social justice will be threatened if the world continues to deepen the gulf between those who have the use of ecological goods and services and those who do not" (TEEB 2008, 25). This euphemistic language evades issues of dispossession, its everyday reality and political-economic drivers.

The TEEB initiative had been sponsored by the Convention on Biodiversity (CBD), whose 2010 Conference of Parties (COP) in Nagoya promoted several TEEB reports on a metaphorically resonant website called the Bank of Natural Capital. At the COP, natural capital valuation was portrayed as a win-win solution for the environment, the economy, and the poor. According to a subsequent UN report, natural capital stocks are "invisible engines of sustainability" (UNEP 2011, ii), as in nineteenth-century metaphors recasting nature as a machine (see, for example, Williams 1980). These multiple metaphors come to frame a specific global agenda: "The development path should maintain, enhance and, where necessary, rebuild natural capital as a critical economic asset and as a source of public benefits" (UNEP 2011, 16).

In the TEEB and UNEP perspectives, natural capital acquires humanlike powers to deliver services, thus anthropomorphically projecting financial assets onto nature: "In more economic terms, it can be said that ecosystem services flow from 'natural capital stocks' (also sometimes termed 'natural assets'), like interest or dividends from the financial stocks" (ten Brink et al. 2012, 5). Why such a metaphor? "We try to use language familiar to people," according to a lead author (interview, Institute for European Environmental Policy [IEEP], February 18, 2015). Indeed, this ontological metaphor has resonance for decision-makers familiar with financial assets and for would-be managers of nature investments.

According to some proponents, though, the valuation of nature may neglect important aspects of natural capital. Payments for ecosystem services (ESS) create risks "where there are collective responsibilities sustaining resources through the commons and/or a culture of stewardship based on responsibility, culture and social norms"; moreover, the natural capital metaphor may give priority to protecting "areas that are more directly used by humans" (ten Brink 2015, 45, from the original English-language version).

Again, according to a lead author of the TEEB report: "Econometrics of natural capital and ecosystem services can open up people to a wider valuation of nature, but there is a risk of closure, seeing only the numbers" (interview, IEEP, February 18, 2015). Indeed, the asset metaphor can render socionatural processes less visible, especially those outside a formal economy. For example, those social relations maintaining nature are reified as relations between things, by analogy with labor productivity being reified as dividends of financial capital assets.

Questioning "Natural Capital": Global Justice Movement

At the Rio+20 summit in 2012, the UNEP agenda featured natural capital as an investment imperative. In particular, the metaphor "natural capital asset" highlighted the source of ecosystem services on which economic activities depend (UNEP 2011, 16). Perhaps unsurprisingly, however, this metaphor provoked suspicion from the global justice movement (BankTrack 2012; People's Summit 2012; No to Biodiversity Offsetting 2013; WDM 2013).

Critics gave several reasons for their suspicions. As a form of nature pricing, natural capital accounting ignores the communities who help to maintain ecosystem services; indeed, such concepts "obscure the social context" of resource flows and usage (Unmüßig et al. 2012, 28). Speaking for many groups at the summit, one NGO elaborated on the commons as a more suitable community defense against global market pressures: "Where markets seek to take power away from the people and distribute resources according to the participants' ability to pay rather than need, a commons-centered approach treats nature, the environment, food, water and other vital aspects of our lives as something we all share rights to and a responsibility for" (WDM 2012). At the 2012 People's Summit, numerous stories emphasized how resistance to enclosure helps communities to defend and develop commons, as a basis for a different global future. "

Such conflicts arose at a World Forum on Natural Capital held in Edinburgh in 2013 and again in 2015. The publicity warned: "With Natural Capital, when we draw down too much stock from our natural environment we also run up a debt which needs to be paid back" (WFNC 2015). Here the creditor is reified as an anonymous thing. The forum's economic metaphor was contrasted with the ecological debt owed to the global South by the North (cf. People's Summit 2012). In both years the Forum was targeted by a North-South global justice network, denouncing the natural capital metaphor. Conference participants "are confusing value with price, and by doing so they open the door for green markets that price everything but value nothing" according to this network (WDM 2013). Others in the network argued that the natural capital metaphor "serves to permit the commodification of nature" (No to Biodiversity Offsetting 2013), and that "It is a con that promotes the interests of businesses in the name of environmental and social protection," representing "corporate polluters seeking to greenwash malpractice" (Open Letter 2015).

The notion of pricing nature empowers financial interests rather than protecting resources, according to a political activist:

> Putting a price tag on nature…will encourage commodification of natural resources and not serve the interest of biodiversity. It will give much control of such resources to corporations and rich members of society. Instead of advocating for market solutions to protect natural resources, we should strengthen local institutions and empower communities. (Teresa Pérez, World Rainforest Movement, quoted in Kenner 2014, 6)

Moreover, the nature valuation process cannot be neutral vis-à-vis the methods, tools, and applications: "The process of valuation is intrinsically linked with the tools that will be used in such valuations. Some argue that the process of valuation is separate from the tool of pricing. However, history clearly shows that the development of the methods is shaped by the tools to be used and vice versa" (Thabit Jacob, co-organizer of the Green Economy in the South conference, Tanzania, 2014, cited in Kenner 2014).

Moreover, argue critics, natural capital depoliticizes the resource issues and power relations: "This economization of nature changes how it is viewed and ultimately undermines political action, which really ought to be committed to public welfare and all nature's functions" (Unmüßig 2014,

12). Natural capital accounting (NCA) helps avoid or displace state responsibility for resource protection: "Good intentions around natural capital are entirely understandable, but reinforcing the idea that everything has a price will not engender the ability to treat the natural world differently" (director of WDM, now Global Justice, quoted in NCI, 2015, 27).

Such global activist networks have opposed NCA for facilitating financial instruments and undermining regulation. They counterpose alternative concepts and metaphors—for example, environmental justice, commons, and a Mother Earth metaphor (Espinosa 2014). With such alternatives, critics have questioned the corporate interest in NCA.

In the form promoted by UN initiatives and business, the natural capital metaphor has divided those claiming to protect the environment. By contrast with the global justice movement, non-governmental conservation organizations (NGCOs) have played a central role, initially in TEEB and later in the Natural Capital Coalition (see next section). Contradictory forms of political engagement have been theorized as a typology. On the one hand, an "embrace" strategy accepts financial metaphors within neoliberal assumptions about NCA. By contrast, a "reject and replace" strategy criticizes those assumptions, while counterposing different metaphors such as organic ones, ecological debt, or Mother Earth (Coffey 2016). This typology was meant to analyze academic perspectives, though it also has relevance to stakeholder agendas (see table 9.1). Let us next trace how NCA has been operationalized.

Strategizing Supply Chains: The Natural Capital Protocol

Natural capital accounting (NCA) has been promoted for and by many business organizations, especially the World Business Council for Sustainable Development (WBCSD). Drawing on the natural capital metaphor, the new TEEB for Business Coalition sought to promote a shift in corporate behavior toward preserving natural and social capital. It formalized earlier collaborations between the WBCSD and non-governmental conservation organizations (NGCOs), such as Conservation International, IUCN, and WWF.

As the coalition warned companies, environmental externalities may jeopardize their economic competitiveness: "Those businesses that fail to adapt ... will lose competitiveness as the value of these resources is realized through tighter regulation, consumer choice and limited supply," warned the director (cited in EY 2014, citing CIMA 2014). This rationale motivated

Table 9.1
Contrary multi-stakeholder alliances vis-à-vis natural capital

Alliance	Natural Capital Coalition (NCC)	Global Justice network
Members	TEEB for Business Coalition, WBCSD, NGCOs (IUCN, WWF, Conservation International), Trucost	Transnational advocacy networks under the slogan global or social justice, e.g., FERN, GJEP, BankTrack, WDM/Global Justice Now, etc.
Nature: ontology	Stocks-assets yielding ecosystem services	Mother Earth protected and maintained by communities
Causes of environmental degradation	Invisibility of nature in economic valuations Vulnerability of natural capital stocks, as the source of ecosystem services	Resource commodification undermining common goods Unequal distribution of land and accumulation of power in a few hands
Fair remedy	Company investment decisions considering all impacts and dependencies on natural capital Anticipate how NC may be internalized via markets and regulation [Silent on development and social equity]	Environmental justice linking commons, communities and resource sovereignty Stronger local economies and territorial rights of communities
Natural-capital accounting (NCA) roles	NCA can help business to manage companies' opportunities and risks (financial and reputational), especially by better managing their supply chains	"Pricing nature" shifts control over resources to powerful financial and corporate interests, rather than protecting biodiversity
Knowledge: valuation methods	Involve NGCOs to standardize NCA methods so that their application will be credible and robust Ensure that NC valuation methods be neutral as regards any application or interpretation	Oppose NCA because the methods are shaped by the tools to be used, especially financial instruments such as ecosystem pricing Share stories of how communities maintain Mother Earth as commons
NC metaphor (Coffey 2015)	Embrace NC for company accounts and supply-chain management	Replace NC with commons and Mother Earth

a new organization to devise and trial a Natural Capital Protocol for supply-chain strategies, as analyzed in this section.

In 2014, the TEEB for Business Coalition was expanded and rebranded as the Natural Capital Coalition (NCC). That change notwithstanding, the earlier problem diagnosis was reiterated: "Economic invisibility has been a major reason for the neglect of natural capital" (NCC 2014, ii). It devised a Natural Capital Protocol which was meant to provide "a comprehensive guide to measuring and valuing natural capital in business decision-making." It promotes the vision of "a world where business conserves and enhances natural capital ... by providing a standardized framework for business to measure and value their direct and indirect impacts (positive and negative) and dependencies on natural capital" (NCC 2015a, 14). The Protocol would start by understanding the business case, impacts and dependencies, risks and opportunities. Such aims set the priorities for identifying relevant activities that benefit from ecosystem services.

This initiative had anticipatory drivers. It frames nature's stock as providing free goods and services, but these "are not typically bartered and sold in the marketplace, so their value is exceedingly hard to price on corporate or government financial statements" (NCC 2014). As a potential driver, more stringent environmental regulation has been anticipated—but not advocated: "The future shock for business is the potential for profit to be wiped out as natural capital is internalized through regulation and markets.... Companies who act now to future-proof are better positioned to manage and thrive in a future 'resource-constrained' world" (Maxwell 2015, 6, 29). Such future anticipation lacks strong regulatory pressures, as one interviewee noted: "Some people see a coherent metrics for natural capital accounting as a driver for improvement. Business people understand the issues around ecosystem services but feel no pressure to internalize the externalities of their operations. Mother Nature calling in her invoice would impose higher financial costs" (interview, industry consultant, January 11, 2016). With this anthropomorphic projection, nature-as-creditor substitutes for regulatory pressures. NCA anticipates future changes in a business's dependence on ecosystem services (ESS), its vulnerability to regulatory changes, and thus options for future strategy. All this warrants more anticipatory management of investment and supply chains.

According to the NCC's first director, business wants to make smarter decisions. Ecosystem services are the Achilles' heel of the economy, so "We must get our ankles covered." By identifying potential impacts on natural capital, the Protocol will help management make the right decisions (van der Gaag 2014). Greater competition for resources "endangers corporate reputations and marketability of products" (CISL 2015, 17). Some companies seek a first-mover advantage in securing their supply chains (NCC director, interview, May 1, 2015).

The Protocol distinguishes between its putatively neutral methods and their specific application by each company. Such assumptions are shared by conservation experts jointly drafting the Protocol: "The method must look with both lenses (society and business) at the same resources, without any moral judgment on choices. The valuation method aims to be neutral regarding any interpretation or application, though neutrality may be difficult to achieve" (interviews, Conservation International members of NCP's Technical Group, June 23, 2015). Likewise, the accounting methods are separable from any subsequent ethical judgments (interview, WWF, June 23, 2015). Indeed, NGCOs see their role as ensuring value-neutral methods.

Despite the putative neutrality of NCA, some insiders emphasize its special relevance to financial instruments. According to a business liaison staff member: "Natural capital valuation could help with the tradable credits already in place. It could help current ones by carrying out a valuation exercise, e.g., of restored wetlands to feed into a wetlands bank, or create potential for a new market" (interview, NCC-WBCSD, June 29, 2015). Likewise, "NCA can potentially play an important role in developing and implementing market-based instruments, such as payment for ecosystem services and biodiversity offset markets," argues a business consultancy (Spurgeon 2014, 6).

Regardless of such schemes, NCC member companies seek methods relevant to their own strategies for investment, supply chains, and product marketing. The Protocol warns companies that their agro-food assets "could become stranded by threats to critical ecosystems" (NCC 2016, 9, citing Oxford University 2013). As a strategic response, "Businesses in the food and beverage sector can use natural capital assessments to inform decisions such as where to grow and invest capital, or withdraw and divest assets, or how to weigh environmental constraints for new or different business models" (NCC 2016, 56). Competitive advantage includes intangible benefits, such as "reputational benefits from own-brand differentiation" (NCC 2016).

Meanwhile a further UNEP report elaborated a strategic rationale for corporate NCA. By linking natural, financial, and reputational assets, environmental stress tests can inform company decisions on investment and supply chains. Such tests would help to avoid "disorderly market responses" to both financial and reputational threats, for example, from "the rise of the civil society divestment movement." For business to deal with such threats, "environmental regeneration will need to be placed within the price system of the real economy" through NCA (UNEP 2015, 15, 4). While "stress" formerly referred to ecosystems, these are newly linked with company reputations and thus resource access—potentially threatened by divestment or boycott campaigns, as well as environmental degradation.

Next, let us survey how the NCA framework relates to practical changes by specific companies.

Turning Nature into an Asset: Three Cases

The Natural Capital Protocol (NCP) has been elaborated and publicized through pilot studies by several companies, which thereby became the NCC's Business Engagement Partners, advised by NGCOs. Although the consequent report remains confidential, the process apparently contributed to those companies' wider CSR strategies. In some cases, the company had been facing attacks on social and/or environmental grounds from NGOs. Each company seeks to protect its reputation and stabilize its natural resource access through environmental stewardship strategies. This section briefly examines three such companies in turn: Kering, Olam, and Coca-Cola. These cases illustrate diverse means and forms of turning nature into an asset.

Kering Group

The Kering Group, encompassing textiles and luxury goods, underwent reputational problems in the early 2000s. Its Puma subsidiary faced NGO protests for causing environmental degradation and labor exploitation. For both issues Puma soon took remedial measures, which became precedents for the entire Kering Group (Baumann-Pauly et al. 2016). For its public relations strategy, one subsidiary funded a film warning against ecological damage, released on World Environment Day, while also cross-promoting environmentally sustainable products (La Redoute 2009). In parallel Kering's Gucci Group decided to eliminate all paper made from Indonesian rainforests and plantations, in partnership with the Rainforest Action Network (Kering 2009).

In 2012, Kering set ambitious environmental targets for 2016, as regards leather traceability, gold sourcing, water pollution, chemical use, and carbon emissions. To guide its efforts, the company developed Environmental Profit and Loss (EP&L) methods. A specific focus was the company's Puma brand of denim products. Its Re-Cut Project redesigns the process "to waste less, recycle more and steer our materials through a more efficient chain."

Puma piloted NCA methods, which were later incorporated into the NCC's framework: "We have now fully integrated this pioneering natural capital accounting tool into our business ... as we explore different options to improve the sustainability of our supply chain." Using such methods, the Group achieved a 10 percent reduction in impact intensity between 2012–2015 (Kering 2015, 3, 7). The methods were extended for comparing the EP&L effects of various supply-chain options, toward its 2025 target for a 40 percent reduction in EP&L, relative to its business growth. Going beyond resource accounting, the company has devised a multi-stakeholder engagement, for instance, for Gucci's cashmere supply.

Over the past couple of decades, cashmere production has been causing environmental degradation and economic insecurity for suppliers. After the end of Mongolia's communist regime in the 1990s, regulations became more lax, export opportunities greatly expanded, and cashmere production increased accordingly, generating a global mass market. As habitual overgrazers, goats depleted grazing areas. For similar reasons, cashmere quality declined, yielding a lower price per unit and so pressurizing herders toward more intensive grazing. The government had abandoned its agricultural extension services, so herders lost any advice on better practices (WCS 2017).

Facing supply-chain difficulties, the company has promoted more sustainable animal husbandry practices and better management of pasture lands (Mehta 2018). "Cashmere supply is the main economic activity in a landscape where we don't own the land" (interview, Kering, February 1, 2019). Its strategy has sought to reconcile economic, social, and environmental goals, as publicized by the Natural Capital Coalition: "Under the Sustainable Cashmere Project, herders receive better or more reliable market prices in return for best practices. They are also afforded more direct market access, and support for improved quality and sustainable, wildlife-friendly grazing practices, pioneered by Wildlife Conservation Society" (NCC 2018, 24).

Herders are given a premium price for cashmere from less-intensive practices, alongside expert support and veterinary care from the WCS.

Environmental improvements have been monitored through ecosystem modeling techniques, provided by a mining company (Hume 2018).

In this case, a multi-stakeholder partnership brought expertise and legitimacy for community-based ecosystem maintenance, illustrating principles of natural capital conservation. This improvement has been turning nature into a shared asset which can more stably and lucratively supply global luxury markets.

Olam International

Olam International, an agro-food conglomerate, became one of the NCC's Business Engagement Partners. It helped develop and test the NCP's Food and Beverage Sector Guide, led by the IUCN. Olam praised the process as helpful for "holistic decision making" to deliver corporate culture change (NCC 2016). Meanwhile the company entered a multi-stakeholder engagement process, responding to protest.

In 2015, Olam had established a joint venture with Gabon's government to establish oil palm plantations (Olam 2015), which were then denounced by NGOs. Although the company had promised to make its operations sustainable, the NGOs argued that "there is still a threat that the plantation project could result in significant deforestation and provoke conflicts over land rights" (Oxfam Australia 2014, 40; also FERN 2016, 11). A protest campaign targeted Olam's oil palm and rubber operations, with a complaint to the Forest Stewardship Council. The environmental damage was further documented by the NGO Mighty Earth (2016); it sought to persuade the world's largest food and agriculture companies "to adopt policies to eliminate deforestation and human rights abuse from their supply chains."

In response to those demands, Olam sought to validate its environmental policy of "Growing Responsibly." It became the first company globally to complete a High Conservation Value assessment under the HCV Resource Network System, and then the first company whose Gabon plantation fulfilled the standards of the Roundtable for Sustainable Palm Oil (RSPO). These activities involved the World Wildlife Fund as a technical partner in certifying Olam's production methods (WWF 2016). Beyond advising the company on compliance, WWF also co-led the RSPO national interpretation process in collaboration with Olam and SIAT, another agribusiness company (WWF 2016). Olam then aimed for 100 percent compliance with the RSPO certification by 2020 (Olam 2016, 6).

Olam also undertook to respect the Central African Forest Initiative (CAFI), which was signed in 2017 by the Gabon government. It undertook to reduce GHG emissions, preserve all High Carbon Stock (HCS) and High Conservation Value (HCV) forests, and cap the amount of deforestation. Olam also signed on to the international convention on "No Deforestation, No Peat, No Exploitation" (NDPE). Olam's undertakings were a change in words rather than practice, according to NGOs such as the World Rainforest Movement, representing social movements (GRAIN and WRM 2017).

In response to such criticism, the company reached an agreement: It suspended any further forest clearing for palm and rubber plantations in its supply chain, while protest was suspended by two international NGOs, Mighty Earth and World Resources Institute. All those parties sought "common ground…on a sustainable and prosperous path forward" (Mighty Earth 2017).

Olam then hosted a visit by international NGOs for discussion with them and a joint platform of twenty civil society groups, "Gabon, Ma Terre, Mon Droit." According to a joint report by the foreign NGCOs and the company, its Social Team "has strong connections to the villages and has previously addressed and documented both the grievances expressed and the solutions proposed" (Olam 2017).

Eventually the company announced its Olam Living Landscapes Policy (OLLP) for a "net-positive" approach for regenerating natural and capital (NCC 2018a, 2018b). This would require "the ongoing support of our partners, including civil society" (Schroeder 2018). Beyond its freeze on deforestation, Olam has been helping Gabon's National Parks Agency set up more parks for nature conservation and ecotourism (Rosner 2018); this presumes a wild nature protected from people.

As the wider political context, any expansion of oil palm monocultures has been opposed by community-based organizations such as Brainforest and Muyissi Environnement. Although welcoming the company's moratorium, Muyissi has seen no environmental improvement in plantation areas, which degrade traditional forest ecosystems: pesticide-based monocultures "lead to an environmental imbalance among plant and insect populations; some species disappear and more powerful pests appear" (WRM and Muyissi 2018). Moreover, such plantations remove or pollute the land necessary for villagers' light economic activities such as food cultivation, which maintain ecosystems (email message, Muyissi, February 2, 2019).

The company established community committees to consult villagers on amenities such as clean water supply. But such consultation has an inequitable basis, given the villagers' difficulty to know long-term consequences, as well as the recent legacy of government repression (interview, Mighty Earth, January 31, 2109).

In sum, Olam was cooperating with international NGCOs for sustainability certification, and it was engaging with communities for a less conflictual basis to source palm oil. Yet its supply chain turns forests into plantations as an asset, undermining and obscuring traditional modes of ecosystem maintenance. Amenities at best play a compensatory role.

Coca-Cola

Coca-Cola has faced significant reputational damage from complicity with death squads (Killer Coke 2004), as well as from large-scale water extraction, drawing attacks for "drinking the world dry" (War on Want 2007). Boycott campaigns became a wake-up call for the company. In response, the company sought to replenish the source of all of its products' water use by 2020. In 2015, the company announced that it had nearly achieved this goal, thus making its production "water neutral" (Kent 2016).

Water replenishment projects aim to enhance its water stewardship role. Their evaluation integrates methods of natural capital and ecosystem services assessment (Denkstatt and Coca-Cola 2016, 8). The company carried out a pilot study of the NCC's Food and Beverage Sector Guide (NCC 2016).

To manage water stress, Coca-Cola devised a Source Water Protection Plan (SWPP), for which "we engage the community, local government, civil society and other businesses to look for ways to collaborate" (Coca-Cola 2016). This program is carried out "with local communities and governments and other respected third-party partners" (Coca-Cola 2015), especially WWF and the Nature Conservancy (WWF 2018). Such NGCOs potentially legitimize the expert methods for identifying multi-stakeholder dependence on natural resources, as a basis for a company to claim water stewardship. Yet its water replenishment substitutes only a small fraction of the water consumed across the company's entire supply chain (MacDonald 2018).

Its water-intensive supply chain has been especially contentious in India. To soften public criticism, the Coca Cola's India Foundation (Anandana) has expanded water replenishment across the country. One means is the "golden triangle"—collaboration between business, government, and community; in some projects Anandana involved local or national NGOs as partners. Its water stewardship program has had "three mantras": providing up-to-task professional resources, including education; empowering small landholders by building water-secure and climate-resilient agricultural capacities; and taking a water-plus approach to raise livelihoods for India's neediest and most water-scarce communities (Coca Cola 2017).

Despite Anandana's efforts at water replenishment, by 2016 at least one-fifth of the company's bottling plants in India were closed—in response to community protests, resource shortages, or orders from India's National Green Tribunal (Down to Earth 2016; India Resource 2016). Since then the company has faced more protest and boycotts, especially in India (Bloomberg 2017). Coca-Cola continues to undergo financial and reputation damage, even supply-chain blockages there. These conflicts arise from a perpetual growth model (Elmore 2015), whose supply chain depends on intensive water extraction. This model turns nature into a non-proprietary asset which can be only somewhat substituted through replenishment activities.

Stewarding Shared Assets, Depoliticizing Resource Conflicts

The business strategies discussed above illustrate some general patterns. In particular, the companies seek to protect access to natural resources which are largely nonproprietary—that is, beyond their own legal ownership or balance sheets, on which the latter depend. Some do so through partnerships with non-governmental conservation organizations (NGCOs) and together they turn natural resources into shared assets, as described in this section.

Global business has sought NGCOs as partners for many environmental initiatives, especially natural capital accounting (NCA). A leading role has been played by the International Union for Conservation of Nature (IUCN). As advertised at IUCN's World Conservation Congress: "Business is increasingly recognized as part of the solution, and NGOs are more than ever willing to discuss and collaborate with business," according to the World Business Council for Sustainable Development (WBCSD 2012). The latter sees an important role for nature conservation groups as "progressive" NGOs:

> NGOs and business have converged in understanding the value of nature for business and for society.... They have had some convergence in the language they use. It's difficult to say that "natural capital" has helped convergence across all

> NGOs, except among the most progressive NGOs, meaning those which want to help business to improve rather than shut the door. (interview, NCC-WBCSD, June 29, 2015)

IUCN has led some pilot studies of the Natural Capital Protocol discussed above. In this role, the IUCN expects that the Natural Capital Protocol will "help businesses understand the risks and opportunities that arise from accounting for natural capital in their decision-making processes" (IUCN 2014). Thus the IUCN projects its social and environmental aims onto business: "We will ensure that the Protocol becomes a valuable and critical tool for the business community to contribute to IUCN's mission of a just world that values and conserves nature" (IUCN 2014), though the concept "justice" appears in no relevant documentation.

The IUCN has had a central role in the Natural Capital Protocol, which elaborates a method for its Business Engagement Partners (BEPs) to identify "natural capital impacts and dependencies across a supply chain" (NCC 2015a). It emphasizes prospects "to unlock hidden value in the supply chain," especially by reframing natural resource issues around "commercial opportunity" and creating "shared value with stakeholders." With such an approach, BEPs can enhance business reputation and thus their social "license to operate" (NCC 2015b).

Indeed, the financial stakes are reputational, given that companies remain vulnerable to public protest. Relative to other environmental approaches: "Natural capital gets more traction with finance departments. At one time, 80 percent of companies' value was tangible—that is, on the books. But now it's only 20 percent; the rest depends on its reputation. Through natural capital accounting, we become aware of negative impacts and how to fix any problems. Otherwise outsiders will push us to do so" (NCC director, interview, May 1, 2015). This warning is reinforced by nature conservation groups: "A company may face blockages from consumer or community action," among other reasons to consider natural capital implications (interview, WWF, June 23, 2015). Such threats and opportunities have been elaborated by several expert reports.

The Trucost consultancy report warns that when companies seek to grow, they encounter several limits and "their ability to achieve revenue targets may be constrained by the potentially rapid, non-linear internalization of natural capital costs through regulation, social campaigns and shortages—and identify alternative strategies to minimize costs and enable growth" (Trucost 2016, 2). A company can turn risks into business opportunities by reconsidering its supply chain through a "resilient sourcing strategy" and "stewardship" interventions (Trucost 2016). By positioning "their business for a low carbon, resource efficient future," they can "demonstrate the shared value they are creating for stakeholders and customers" (Trucost 2016, 12). According to the report's lead author, business interest in natural capital has several motives and aims. Going beyond previous methods, NCA brings all natural resources within a common framework, toward a holistic long-term view. The accounting methods focus on tangible but nonproprietary assets; these are more at risk and less under a company's control than proprietary ones (interview, Trucost, August 26, 2016).

NCA helps a company's sustainability unit obtain engagement from its finance unit, for example by monetizing resource dependencies or by signaling potential jeopardy of its social license to operate. Such methods have gained much interest from companies in resource hot spots, especially societal conflicts around water, such as Coca-Cola in India or the Columbian coffee industry after the long civil war there. Thus NCA can inform company strategies to gain a stewardship role by restructuring its supply chains and engaging community representatives (interview, Trucost, August 26, 2016).

Working with IUCN experts, a Cambridge University program has been asking companies how NCA could inform their strategies. Rather than focus on one resource, a company intervention could "collectively manage water, soil and biodiversity" (CISL 2015, 28). Along those lines, business stakeholders had shifted their focus "towards a more forward-looking, opportunistic approach when assessing natural resource challenges.... The inclusion of these concerns in business decision-making is now considered as a financial and market opportunity by business" (CISL 2016, 6). This implies a holistic basis for a company stewardship strategy to gain legitimacy.

For such opportunities, accounting methods describe natural resources through a universal equivalence, thus homogenizing them, as in a joint academic report with the Kering Group. Here all environmental effects are quantified as "profit and loss"; biodiversity becomes a "portfolio" which confers resilience on ecosystems; and, in the name of contextualizing environmental issues, other stakeholders are relegated to casualties of potential damage or scarcity (CISL-Kering-NCP 2016).

Nature accounting likewise turns water into homogeneous quantities that can be substituted and replenished. These remain separate from community maintenance and its cultural meanings, which are rendered invisible (as in the Coca-Cola case above). In those ways, a socionatural process is reified as an a-social asset, whose nonproprietary "portfolio" warrants a holistic responsible investment manager, as if the asset were proprietary.

By contrast to that prevalent nature/society binary, one TEEB report highlighted social tensions inherent in maintaining and accounting for natural capital. "If valuing biological resources is a tool to improve *in situ* conservation, it assumes that local stakeholders have sufficient incentives to maintain a given ecosystem against other competing uses," so contradictory drivers warrant attention and caution. Yet nature accounting readily separates people from nature by "simplifying its meaning and value to human societies" (Brondízio and Gatzweiler 2010, 19, 28).

This caveat has been echoed by some NGCO experts. For example, communities routinely maintain natural capital, so "the valuation methods should make their labor visible, especially in a local context, though the relations can get buried in multiple metrics" (interviews, Conservation International members of NCP's Technical Group, June 23, 2015). Indeed, ecosystem services are generally attributed to natural assets, while communities protecting commons readily become invisible. Likewise, the analogy with financial capital has limitations for fairness issues: "The capital metaphor has the disadvantage that it cannot encompass shared resources, and sometimes capital is destroyed.... There is a justice challenge if multinational companies value natural capital only for its dividends, at the expense of indigenous people and small businesses" (interview, NCC ex-director, March 10, 2015). Indeed, such conflicts arise pervasively around business claims and strategies for shared resources. Yet NCA guidance implies that corporate investment in natural capital can assess diverse services as comparable, readily protect ecosystems, and accommodate their multi-stakeholder dependents. On this basis, NCA guidance envisages consensual win-win scenarios for companies stewarding nonproprietary shared assets as if they were proprietary.

Conclusion: Turning Nature into an Asset, Reifying Socionatural Processes

Coming from development theorists, "natural capital" originally meant a socionatural resource empowering community development for better environments and livelihoods (Wilshusen 2014). Dominant institutions later took up the metaphor "natural capital *asset*" for evaluating and protecting the source of ecosystem services. This metaphor reifies socionatural processes as a thing-like a-social asset providing ecosystem services for market-based economic activities.

In the imaginary of capitalist environmentalism, managers holistically govern mutual relations between nature and society; in practice, they construct such a nature/society binary for specific production aims (see, for example, Lohmann 2019). Ecosystem services become analogous to dividends, which are likewise reified as an inherent property of finance capital. Here "natural capital asset" serves as an ontological metaphor (cf. Lakoff and Johnson 1980). It extends a long history of conferring capitalist metaphors on nature, thus naturalizing specific forms of private appropriation (Williams 1980).

Underlying the asset metaphor is a problem diagnosis that ecosystem degradation results from the economic invisibility of nature. As the remedy, therefore, it must be made visible through natural capital assessment or accounting (NCA), which has various institutional contexts. Corporate NCA has a more specific driver: a company's access to natural resources faces several threats. These could jeopardize the basic conditions of a company's production process, by analogy with restrictions on GHG emissions (see, for example, Felli 2014).

To address those threats, NCA methods have been elaborated by a partnership between business organizations, non-governmental conservation organizations (NGCOs) and other experts. Their joint Natural Capital Protocol provides "a comprehensive guide to measuring and valuing natural capital in business decision-making." NCA evaluates how a business depends on ecosystem services—especially by identifying biophysical, financial, and reputational risks. Such accounting plays a performative role in attributing value to entities (see Muniesa et al. 2017).

Beyond a company's own resource demands, NCA can identify multi-stakeholder dependence on shared assets. In practice, especially in the global South, ecosystems are generally maintained (or transformed) by everyday social-communal labor, involving distinctive cultural meanings. This socionatural process becomes reified as inherent properties of natural capital assets, though some NGCO experts highlight their maintenance by communities. NCA informs corporate strategies for managing supply chains, stakeholder engagement, and reputations.

Such strategies turn nature (or ex-nature) into a more effective asset, functioning as if it were proprietary, even if outside a company's ownership.

Each strategy remains contingent on specific actors, contexts, and global markets—for example, Kering-Gucci facilitates less-intensive animal husbandry, while making visible the herders'" role in ecosystem maintenance as shared value for a luxury product. Olam turns forests into monoculture plantations-as-assets, distinguished from forests as non-human nature, meanwhile keeping invisible and undermining villagers' role in ecosystem maintenance. Cola-Cola seeks to stabilize water sources as an asset, amidst rival burdens on state-run water services; the company implements conservation and replenishment initiatives, which hardly compensate for the significant depletion by its water-intensive supply chain.

These strategies extend the long-standing process of neoliberalizing nature—"off-loading responsibilities to the private sector and/or civil society groups" (Castree 2008). This process depoliticizes mutual dependencies, societal conflicts, and power inequalities around natural resources (see, for example, Holmes and Cavanagh 2016; Unmüßig 2014). Those efforts can help incorporate community activities or representatives into a company's economic logic and development model (as in nature conservation initiatives; see Holmes and Cavanagh 2016). These strategies facilitate rent-seeking—that is, a company's favorable access to natural resources on which its supply-chain depends.

Acknowledgments

Helpful editorial comments were received from Romain Felli, Jutta Kill, Larry Lohmann, and the editors of this book, Kean Birch and Fabian Muniesa.

References

Adaman, F., and Madra, Y. 2014. Understanding Neoliberalism as Economization: The Case of the Environment. In *Global Economic Crisis and the Politics of Diversity*, edited by Y. Atasoy, 29–51. London: Palgrave Macmillan.

Akerman, M. 2005. What Does "Natural Capital" Do? The Role of Metaphor in Economic Understanding of the Environment. *Environmental Education Research* 11 (1): 37–52.

Andonova, L. B., and Hoffmann, M. J. 2012. From Rio to Rio and Beyond: Innovation in Global Environmental Governance. *Journal of Environment & Development* 21 (1): 57–61.

BankTrack. 2012. Position on the Natural Capital Declaration, http://www.banktrack.org/show/news/banktrack_position_on_the_natural_capital_declaration.

Baumann-Pauly, D., Scherer, A. G., and Palazzo, G. 2016. Managing Institutional Complexity: A Longitudinal Study of Legitimacy Strategies at a Sportswear Brand Company. *Journal of Business Ethics* 137: 31–51.

Birch, K. 2017. Financing Technoscience: Finance, Assetization and Rentiership. In *The Routledge Handbook of the Political Economy of Science*, edited by D. Tyfield, R. Lave, S. Randalls, and C. Thorpe, 169–181. London: Routledge.

Birch, K. 2019. Technoscience Rent: Toward a Theory of Rentiership for Technoscientific Capitalism. *Science, Technology and Human Values*, doi:10.1177/0162243919829567.

Bloomberg. 2017. Pepsico, Coca-Cola Fight Patriotism in Drought-Hit Kerala, March 16, https://economictimes.indiatimes.com/industry/cons-products/food/pepsico-coca-cola-fight-patriotism-in-drought-hit-kerala/articleshow/57665983.cms.

Brondízio, E. S., and Gatzweiler, F. W. 2010. Socio-cultural Context of Ecosystem and Biodiversity Valuation. In *The Economics of Ecosystems and Biodiversity Ecological and Economic Foundations*, edited by P. Kumar, 50–181. London: Earthscan.

Castree, N. 2008. Neoliberalising Nature: The Logics of Deregulation and Reregulation. *Environment and Planning A* 40: 131–152.

CIMA. 2014. *Accounting for Natural Capital: The Elephant in the Boardroom*. London: Chartered Institute of Management Accountants, https://www.cimaglobal.com/Documents/Thought_leadership_docs/Sustainability%20and%20Climate%20Change/CIMA-accounting-for-natural-capital.pdf.

CISL. 2015. *Doing Business with Nature: Opportunities from Natural Capital*. Report by the Natural Capital Leaders Platform, Cambridge Institute for Sustainability Leadership (CISL), https://www.cisl.cam.ac.uk/resources/publication-pdfs/doing-business-with-nature.pdf.

CISL. 2016. *Operationalising Natural Capital: Managing Opportunities and Risks from Natural Resources*. Cambridge, UK: Cambridge Institute for Sustainability Leadership.

CISL-Kering-NCP. 2016. *Biodiversity and Ecosystem Services in Corporate Natural Capital Accounting: Synthesis Report*. Cambridge, UK: Cambridge Institute for Sustainability Leadership.

Coca-Cola. 2015. Collaborating to Replenish the Water We Use, http://www.coca-colacompany.com/stories/collaborating-to-replenish-the-water-we-use.

Coca-Cola. 2016. Mitigating Water Risk for Communities and for Our System, http://www.coca-colacompany.com/stories/mitigating-water-risk-for-communities-and-for-our-system.

Coca-Cola. 2017. Vision for a Water Positive India: Coca-Cola India's Three Mantras on Reinforcing Water Stewardship, August 28, 2017, https://www.coca-colaindia.com/point-of-view/vision-for-a-water-positive-india-coca-cola-india-s-three-mantras-on-reinforcing-water-stewardship.

Coffey, B. 2016. Unpacking the Politics of Natural Capital and Economic Metaphors in Environmental Policy Discourse. *Environmental Politics* 25 (2): 203–222.

Corson, C. 2010. Shifting Environmental Governance in a Neoliberal World: US AID for Conservation. *Antipode* 42 (3): 576–602.

Darwin, C. 1859. *On the Origin of Species by Means of Natural Selection*. London: Murray.

Denkstatt and Coca-Cola Company. 2016. *Natural Capital Accounting: The Coca-Cola Water Replenishment Program*, http://naturalcapitalcoalition.org/wp-content/uploads/2016/07/Denkstatt_Natural_Capital_Accounting.pdf.

Dewey, J. 1939. *Theory of Valuation*. Chicago: University of Chicago Press.

Down to Earth. 2016. Coca-Cola Has Closed 20 Per Cent of Its Bottling Plants in India: Report, https://www.downtoearth.org.in/news/environment/coca-cola-has-closed-20-per-cent-of-its-bottling-plants-in-india-report-53273.

Elmore, B. E. 2015. *Citizen Coke: The Making of Coca-Cola Capitalism*. New York: Norton.

Espinosa, C. 2014. The Advocacy of the Previously Inconceivable: A Discourse Analysis of the Universal Declaration of the Rights of Mother Earth at Rio+20. *Journal of Environment & Development* 23: 391–416.

EY. 2014. Natural Capital Depletion Will Leave Businesses Dangerously Exposed, Warns New Report,

Fairhead, J., Leach, M., and Scoones, I. 2012. Green Grabbing: A New Appropriation of Nature? *Journal of Peasant Studies* 39 (2): 237–261.

Felli, R. 2014. Climate Rent. *Historical Materialism* 22: 251–280.

Felli, R. 2015. Environment, Not Planning: The Neoliberal Depoliticisation of Environmental Policy by Means of Emissions Trading. *Environmental Politics* 24 (5): 641–660.

FERN. 2016. *Financing Land Grabs and Deforestation: The Role of EU Banks and Investors*. Brussels: FERN.

FoEE. 2012. Pursuit of the Green Economy in the Rio+20 debate, http://www.foeeurope.org/green-economy-rio-conference.

Forsyth, T., and Sikor, T. 2013. Forests, Development and the Globalisation of Justice. *Geographical Journal* 179: 114–121.

Gómez-Baggethun, E., and Naredo, J. M. 2015. In Search of Lost Time: The Rise and Fall of Limits to Growth in International Sustainability Policy. *Sustainability Science* 10: 385–395.

GRAIN & WRM. 2017. L'escroquerie des plantations de palmiers à huile "verts": le cas d'Olam, https://www.grain.org/fr/bulletin_board/entries/5715-l-escroquerie-des-plantations-de-palmiers-a-huile-verts-le-cas-d-olam.

Haeckel, E. 1866. *Generelle morphologie der organismen*. Berlin: Reimer.

Holmes, G., and Cavanagh, C. J. 2016. A Review of the Social Impacts of Neoliberal Conservation: Formations, Inequalities, Contestations. *Geoforum* 75: 199–209.

Hume, M. 2018. The Wild Best. *AFR* magazine, April: 40–44, https://www.riotinto.com/documents/afr_mag_mongolia.pdf.

India Resource. 2016. Campaign to Hold Coca-Cola Accountable, http://www.indiaresource.org/campaigns/coke/.

IUCN. 2014. IUCN Leads the Way towards Valuing Nature in Business, International Union for Conservation of Nature, https://www.iucn.org/content/iucn-leads-way-towards-valuing-nature-business.

Kenner, D. 2014. Who Should Value Nature? http://whygreeneconomy.org/wp-content/uploads/2014/12/Who-should-value-nature.-Why-Green-Economy.pdf.

Kent, M. 2016. Our Water Wake-Up Call; What Will Be Yours? [CEO, Coca-Cola], http://naturalcapitalcoalition.org/our-water-wake-up-call-what-will-be-yours/.

Kering Group. 2009. Gucci Group Commits to Protecting Indonesia's Rainforests, https://www.kering.com/en/news/gucci-group-commits-protecting-indonesia-rainforests.

Kering Group. 2015. *Environmental Profit & Loss (E&PL): 2015 Group Results,*

Kill, J. 2016. *The Kasigau Corridor REDD+ Project in Kenya: A Crash Dive for Althelia Climate Fund*. Re:Common, www.recommon.org and Counter Balance, www.counter-balance.org.

Killer Coke. 2004 (start date). http://killercoke.org.

Laing, T., Taschini, L., and Palmer, C. 2015. *Understanding the Demand for REDD+ Credits*. Centre for Climate Change Economics and Policy Working Paper No. 218. Grantham Research Institute on Climate Change and the Environment Working Paper No. 193.

La Redoute. 2009. La Redoute soutient le film "Home," https://www.kering.com/cn/news/redoute-soutient-film-home.

Lohmann, L. 2019. Natures, Rights and Political Movements. In *The Right to Nature: Social Movements, Environmental Justice and Neoliberal Natures*, edited by E. Apostolopoulou and J. A. Cortes-Vazquez, 234–247. London: Earthscan/Routledge.

MacDonald, C. 2018. Coke Claims to Give Back as Much Water as It Uses: An Investigation Shows It Isn't Even Close. *The Verge*, May 18, https://www.theverge.com/2018/5/31/17377964/coca-cola-water-sustainability-recycling-controversy-investigation.

MacDonald, K. I. 2010. The Devil Is in the (Bio)diversity: Private Sector "Engagement" and the Restructuring of Biodiversity Conservation. *Antipode* 42 (3): 513–550.

MacDonald, K. I., and Corson, E. 2012. "TEEB Begins Now": A Virtual Moment in the Production of Natural Capital. *Development and Change* 43 (1): 159–184.

Marx, K. 1862. Letter to Engels, June 18, 1862. In *K. Marx and F. Engels, Selected Correspondence*. Moscow: Foreign Languages Publishing House.

Maxwell, D. 2015. *Valuing Natural Capital: Future Proofing Business and Finance*. Austin, TX: Greenleaf.

MEA (Millennium Ecosystem Assessment). 2005. *Ecosystems and Human Well-Being: Synthesis*. Washington, DC: Island Press.

Mehta, A. 2018. Natural Capital Case Study: Kering Gets to the Bottom of Its Supply Chain with EP&L. *Ethical Corporation*, http://www.ethicalcorp.com/natural-capital-case-study-kering-gets-bottom-its-supply-chain-epl.

Mighty Earth. 2016. *Palm Oil's Black Box: How Agribusiness Giant Olam's Emergence as a Major Palm Oil Trader Is Putting Forests in Southeast Asia and Gabon at Risk*. Washington, DC: Mighty Earth/Brainforest.

Mighty Earth. 2017. Olam and Mighty Earth Agree to Collaborate on Forest Conservation and Sustainable Agriculture in Highly Forested Countries, February 21, http://www.mightyearth.org/olam-and-mighty-earth-agree-to-collaborate/.

Muniesa, F. 2012. A Flank Movement in the Understanding of valuation. *Sociological Review* 59 (s2): 24–37.

Muniesa, F., Doganova, L., Ortiz, H. Paterson, F., Bourgoin, A., Ehrenstein, V., Juven, P.-A., Pontille, D., Saraç-Lesavre, B., and Yon, G. 2017. *Capitalization: A Cultural Guide*. Paris: Presse des Mines.

NCC. 2014. *Valuing Natural Capital in Business. Taking Stock: Existing Initiatives and Applications*. Natural Capital Coalition, http://naturalcapitalcoalition.org/.

NCC. 2015a. *Developing the Natural Capital Protocol and Sector Guides for Business*.

NCC. 2015b. *The Natural Capital Protocol: Feedback Report from Business Engagement Partner Interviews*.

NCC. 2016. *Natural Capital Protocol: Food and Beverage Sector Guide*, http://naturalcapitalcoalition.org/wp-content/uploads/2016/07/NCC_FoodAndBeverage_WEB_2016-07-12.pdf.

NCC. 2018a. Coalition Organization Olam Announces New Approach to "Creation & Restoration of Natural & Social Capital," https://naturalcapitalcoalition.org/coalition-organization-olam-announces-new-approach-to-creation-restoration-of-natural-social-capital/.

NCC. 2018b. *This Is Natural Capital*, https://naturalcapitalcoalition.org/this-is-natural-capital-2018/.

NCI. 2015. *Natural Capital Initiative 2015: Valuing Our Life Support Systems 2014. Summit Summary Report*, http://www.naturalcapitalinitiative.org.uk/wp-content/uploads/2015/06/Valuing-our-Life-Support-Systems-summit-report.pdf.

No to Biodiversity Offsetting! 2013. http://no-biodiversity-offsets.makenoise.org/.

Olam. 2015. Sale and Lease-Back of Plantation and Milling Assets in Awala for US$130m, https://www.olamgroup.com/news/all-news/press-release/olam-palm-gabon-enters-sale-long-term-lease-rights-land-sale-lease-back-plantation-milling-assets-awala-us130m.html.

Olam. 2016. *Corporate Responsibility and Sustainability Report 2015*, http://olamgroup.com/wp-content/uploads/2014/02/16395-Sustainability-Report-2016-Exec-summary_V17_20p_lowres.pdf.

Olam. 2017. Olam Hosts Mighty Earth in Gabon, https://www.olamgroup.com/news/all-news/news-bites/olam-hosts-mighty-earth-gabon.html.

Open Letter. 2015. Natural Capital or Corporate Greenwash?. Open Letter on World Forum on Natural Capital. November 23, Edinburgh, http://bellacaledonia.org.uk/2015/11/23/open-letter-on-world-forum-on-natural-capital/.

Oxfam Australia. 2014. *Banking on Shaky Ground: Australia's Big Four Banks and Land Grabs*. Carlton, Victoria: Oxfam Australia.

Oxford University. 2013. *Stranded Assets in Agriculture: Protecting Value from Environment-Related Risks*. Stranded Assets Programme, Smith School of Enterprise and the Environment, http://www.smithschool.ox.ac.uk/research-programmes/stranded-assets/Stranded%20Assets%20Agriculture%20Report%20Final.pdf.

Pearce, D., Markandya, A., and Barbier, E. 1989. *Blueprint for a Green Economy*. London: Earthscan.

People's Summit. 2012. Final Declaration, People's Summit at Rio+20 for Social and Environmental Justice in Defence of the Commons, Against the Commodification of Life, https://globalforestcoalition.org/wp-content/uploads/2012/07/PeoplesSummit-FinalDeclaration-ENG.pdf.

Porritt, J. 2006. *Capitalism as if the World Matters*. London: Earthscan.

Robertson, M. M. 2006. The Nature that Capital Can See: Science, State, and Market in the Commodification of Ecosystem Services. *Environment and Planning D: Society and Space* 24: 367–387.

Robinson, J. G. 2012. Common and Conflicting Interests in the Engagements between Conservation Organizations and Corporations. *Conservation Biology* 26 (6): 967–977.

Rosner, H. 2018. Palm Oil Is Unavoidable. Can It Be Sustainable? *National Geographic*, December, https://www.nationalgeographic.com/magazine/2018/12/palm-oil-products-borneo-africa-environment-impact/.

Schroeder, E. 2018. Olam Introduces New Sustainability Policy. *World Grain*, June 4, https://www.world-grain.com/articles/9648-olam-introduces-new-sustainability-policy.

Scott, J. C. 1998. *Seeing like a State: How Certain Schemes to Improve the Human Condition Have Failed*. New Haven, CT: Yale University Press.

Spurgeon, J. P. G. 2014. *Comparing Natural Capital Accounting Approaches, Data Availability and Data Requirements for Businesses, Governments and Financial Institutions: A Preliminary Overview*. Final report to the EU Business and Biodiversity Platform, performed under the ICF contract.

Sullivan, S. 2013. Banking Nature? The Spectacular Financialisation of Environmental Conservation. *Antipode* 45 (1): 198–217.

TEEB. 2008. *The Economics of Ecosystems and Biodiversity: An Interim Report*, edited by Pavan Sukhdev, Study Leader. Brussels: European Commission, http://ec.europa.eu/environment/nature/biodiversity/economics/pdf/teeb_report.pdf.

ten Brink, P. 2015. What Is Natural Capital? In *Nature et richesse des nations*, 41–51. Paris: Service de l'Économie, de l'Évaluation et de l'Intégration du Développement Durable (SEEIDD) du Commissariat Général au Développement Durable (CGDD), Gouvernement de France.

ten Brink, P., Mazza, L., Badura, T., Kettunen, M., and Withana, S. 2012. *Nature and Its Role in the Transition to a Green Economy*, TEEB, http://www.teebweb.org/wp-content/uploads/2013/04/Nature-Green-Economy-Full-Report.pdf.

Trucost. 2013. *Natural Capital at Risk: The Top 100 Externalities of Business*. Report for TEEB for Business Coalition, http://www.naturalcapitalcoalition.org/js/plugins/filemanager/files/TEEB_Final_Report_v5.pdf.

Trucost. 2016. *Growing Business Value in an Environmentally Challenged Economy*. London: Trucost.

UNEP. 2011. *Towards a Green Economy: Pathways to Sustainable Development and Poverty Eradication*. Nairobi: UN Environment Programme.

UNEP. 2015. *Pathways to Scale: Aligning the Financial System with Sustainable Development*. Nairobi: UN Environment Programme.

Unmüßig, B. 2014. *On the Value of Nature: The Merits and Perils of a New Economy of Nature*. Berlin: Heinrich Böll Foundation.

Unmüßig, B., Sachs, W., and Fatheuer, T. 2012. *Critique of the Green Economy*. Brussels: Heinrich Böll Foundation.

van der Gaag, P. 2014. Protecting the Achilles' Heel of Our Economy. Speech at Natural Capital Initiative (NCI) Conference, November, www.youtube.com/watch?v=U4W0w31r_cw.

War on Want. 2007. Coca-Cola: Drinking the World Dry, http://www.waronwant.org/media/coca-cola-drinking-world-dry.

WBCSD. 2012. *Key highlights: WBCSD at IUCN's World Conservation Congress*. World Business Council for Sustainable Development. Jeju, South Korea, September 6–15.

WCS. 2017. Sustainable Cashmere Stakeholder Debate: Stakeholder Viewpoint Handout. New York: World Conservation Society, Education Programme, http://bronxzoo.com/programs.

WDM. 2012. *Rio+20: Whose Green Economy?* London: World Development Movement [renamed Global Justice Now].

WDM. 2013. World Forum on Natural Capital: The Great Nature Sale, http://www.globaljustice.org.uk/sites/default/files/files/resources/nature_sale_briefing.pdf

WFNC. 2015. World Forum on Natural Capital, https://naturalcapitalforum.com/news/archive/december-2015/.

Williams, R. 1980. *Problems in Materialism and Culture*. London: Verso.

Williams, R. 1983. *Keywords: A Vocabulary of Culture and Society*. Oxford: Oxford University Press.

Wilshusen, P. R. 2014. Capitalising Conservation/Development: Dissimulation, Misrecognition and the Erasure of Power. In *Environmental Conservation in the Neoliberal Age*, edited by B. Büscher, W. Dressler, and R. Fletcher, 127–157. Tucson, AZ: University of Arizona Press.

WRM and Muyissi Environnement. 2018. Gabon: OLAM's Industrial Oil Palm Plantations Deprive Community of Sanga of Access to Safe Water. *WRM Bulletin* 240, https://wrm.org.uy/bulletins/issue-240.

WWF. 2016. Strong Message that Palm Oil Companies Can Expand Responsibly. Washington, DC: World Wildlife Fund, http://wwf.panda.org/?284530/WWF-welcomes-first-RSPO-certification-of-a-new-palm-oil-development-in-Africa.

WWF. 2018. WWF and the Coca-Cola Company Have Worked Together. October 18, https://www.wwf.org.uk/who-we-are/who-we-work-with/coca-cola.

IV Turning Publics into Assets

10 English Higher Education: From a Public Good to a Public Asset

Sveta Milyaeva and Daniel Neyland

Introduction

The introduction of university tuition fees has been a contentious issue in UK politics for the past decade. As well as igniting wide public debate, it also gave rise to fierce student protests and contributed to a decline in support for the Liberal Democrat Party (Lewis et al. 2010; Phipps 2014; Gil 2015). This social unrest and change in the political landscape are one outcome of the profound transformation of the way English higher education is funded. It used to be regarded as a "public good," funded through a block grant that allowed English universities to provide education at a very low cost. But through gradual withdrawal of central government funding, the provision of undergraduate university teaching is now financed by tuition fees covered by government loans. As a result, higher education has been transformed from a public good to a public asset.

In this chapter, we focus on the reorganization of government funding for higher education that has enabled this creation of a public asset, what this entails, and the consequences that follow. We analyze this transformation in order to understand the specific features of financialization (Chiapello 2015; Engelen 2008; Van Der Zwan 2014; Davis and Kim 2015) that transform the public sector. The financialization of services that used to be considered the responsibility of the state has attracted little sociological attention and so here we will shed light on the practices of valuation to see how the monetary value of both the public good of higher education and the student loans are performed and accounted for.

The chapter starts by exploring the notion of a public good as understood by economists and how this has been analyzed and critiqued by sociologists and scholars from science and technology studies (STS). We then focus on recent STS work on assetization to open up a space for considering the creation of public assets. We use the empirical focus of higher education funding in England and its transformation from direct funding to income contingent repayment loans to explore in detail how a public asset is made. We argue that what is particularly important is the creation of an impaired asset through accounting techniques that continuously (re) compose its value. Impairment, we conclude, is what enables the asset to maintain the principal features of a public good even when transformed into a public asset.

Theorizing a Public Good and a Financial Asset

The provision of public education was first formally classified in economics as a public good by Paul Samuelson (1955), a Nobel Prize laureate. It is said that Samuelson was an economist who "transform[ed] his discipline from one that ruminates about economic issues to one that solves problems" (Weinstein 2009) and was noted for making public good a textbook category. According to economics, public goods have a set of particular characteristics. A public good is non-rivalrous, it is a "collective consumption good," which means it can be used simultaneously by more than one individual without undermining the quality of the good (i.e., all the individuals using the good would benefit from its consumption in equal measure) (Samuelson 1954). A public good can also be consumed by everyone and not just those who paid for it (i.e., it is non-excludable). As a result of these features, economists suggest private (commercial, market) producers are prevented from benefiting from the provision of public goods to consumers, resulting in their underproduction, prompting governments to enter the marketplace and become the provider of public goods.[1]

Publicly funded universities are routinely praised for delivering economic and social benefits (OECD 2015; Mountford-Zimdars et al. 2013; AAAS 2016) fitting a broader, normative sense of public good. However, transforming higher public education into a governmental asset is a relatively new development. We start by looking more closely at the notion of a public good before paying close attention to the practices of economic *valuation* which we suggest underpin the transformation of such goods into assets.

Callon (1994) rejects the notion that science can be considered a public good through being non-rivalrous and non-excludable. Instead he argues that scientific knowledge is as rivalrous and excludable as any other good. Science in this view accomplishes its status as a public good through "hybrid collectives" that continuously produce variety that leads to social change (Callon 1994, 407). Alternatively, Mirowski (2011) discusses public goods through the lens of its origin and evolution within neoliberal economics. He suggests that the concept has been contingent on the justification of government military spending changing attitudes to the public funding of scientific knowledge and resulting in "the now-pervasive habit of treating the genesis of scientific knowledge as if it were production of a 'thing,' on a par with any other commodity" (Mirowski 2011, 58). Despite these differences in ways of critiquing the notion of public good, one shared premise of these authors is that science attains its status as a public good as a result of overly simplistic assumptions regarding the production of scientific knowledge.

Recently the range of public goods discussed has been broadened to include sustainability regulations and government cultural policies (Doganova and Laurent 2016; Pallesen 2016; O'Brien 2016). Here variation in public goods is linked together not only by the specific public nature of these goods but also by the focus of analysis, namely the practices of economic valuation (Roscoe and Townley 2016). The process of valuation is of an instrumental importance to these accounts of negotiating the worth of public goods, and this choice of an analytical tool is grounded in the turn of sociological attention to economic value and its formation (Stark 2011; Helgesson and Muniesa 2013; Kornberger et al. 2015).

Although this move to study the practices of valuation that underpin the constitution of public goods is appealing, what of the move to transform such goods into assets? The empirical case considered here is how English higher education funding underwent a transformation from direct public financing (a public good) to funding through the provision of loans to students; since loans are considered to be *financial* assets, here we focus on this type of asset being fully aware that an asset could be anything that enables capitalization. A financial asset is a category of financial accounting that is defined as "a resource controlled by the entity as a result of past events and from which future economic benefits are expected to flow to the entity" (IASB 2015, 220). Accounting for this resource thus requires stating a financial position that includes not just its costs but also the likelihood of accruing future financial benefits from an asset. In this way, financial assets, as an income-generating resource, produce capital, and can be viewed as vehicles for capitalization. The latter is regarded as "the present value of a future stream of earnings: it tells us how much a capitalist would be prepared to pay *now* to receive a flow of money *later*" (Nitzan and Bichler 2009, 153, emphasis in original).

Research on the practices of valuation can be applied to these forms of assets in a similar manner to the aforementioned treatment of goods. Hence emerging STS research on the practices of valuation of future income streams—or assetization—focuses on practices that "settle" the value of "unsettled" assets, be it a financial value that accommodates moral and political values (Ortiz 2013) or a business model as an assetization device (Doganova and Muniesa 2015). One of the most immediate and primary practices are accountancy techniques, or "the miracles of bookkeeping" (Quinn 2017). Placing accounting practices at the heart of constructing profits and losses (or assets and expenses) and stressing that "'profits' are, quite literally, constructed by accountants," Hatherly et al. (2008) emphasize the importance of up-close study of financial reporting in the making (e.g., how accounting and bookkeeping training affects practices of classification and concept-application). This importance is further emphasized by Menniken and Millo (2016) in their study of the development of UK impairment rules for asset accounting wherein the authors demonstrate how such rules are rooted in different forms of valuation that result in the "hybridization" of managerial knowledge and financial economics. Intangible asset accounting is also responsible for theorizing business entities (e.g., a firm) (Birch and Tyfield 2013; Birch 2017).

Collectively, these studies of goods and assets share an analytical focus on valuation practices that moves away from treating value as a stable, objective, exogenous entity that exists independent of practices of valuation. Instead, following Dewey (1915, 1939), Muniesa (2012, 28, 24) suggests a "pragmatics of valuation in finance," that assumes no "distance between value and its measure." In this way, financial value is a *practice* (an act) and financial valuation involves double acts of capitalization involving both "the appraisal of the characteristics of something in terms of its value and the setting that thing for the purpose of making it valuable" (Muniesa 2011, 31).

How, then, might we take on this focus on valuation practices, accounting, and double acts of capitalization in making sense of the switch from public goods to public assets in English higher education? We suggest that one means to do this is provided by an expansion of the pragmatics of valuation. Of crucial importance here is how acts of valuation relate to what is valued. To understand the process of assetization it is important to see value as action, a social practice, and process of doing (Muniesa 2012; Birch 2017) since "value both as an idea and as existence depends upon judgement on what to do" (Dewey 1915, cited in Muniesa 2012, 26). But what is also important, we argue, is to pay attention to what to do with what—the "thing" that has been made valuable that appears in the theoretical account of valuation (Muniesa 2012; Birch 2016).

Here we can return to Dewey (1937). He rigorously analyzed how a matter "is subjected to controlled inquiry," how judgments are made about endlessly diverse subjects. Dewey (1937, 102) argued that in making a judgment about something—in our case committing an act of valuation—"formal conceptions arise out of the ordinary transactions; they are not imposed upon [the transactions] from on high or from any external and *a priori* source." This happens through inquiry that turns "indeterminate situations" (what is the value of something?) into "determinate" situations (the value is this) and such transformation is not just a matter of doubt, a mental task. "The doubtful [does not entirely] belong to us," it belongs to the indeterminate situation (which involves the composition of the act of valuation, but also the matter that is subjected to the valuation). It becomes determinate or conceptualized "only by operations which ... modify *existing* conditions" (Dewey 1937, 106, emphasis added). The matter under valuation cannot be disregarded, it is central to the valuation and, subsequently, its value. In the case of public assets what we need to understand, then, is the specific acts through which accounting becomes a basis for constituting in our case higher education and associated student loans as an asset. In treating the transformation of public goods into public assets as a valuation practice, we thus need to account for the *kinds* of entities under assessment—the form that goods, commodities, and assets take—and the valuation practices that accomplish those forms (see Braun, this volume).

English Higher Education and Its Funding

How does a public good become a public asset? In what way is the anomalous nature of a public good consequential to the form of asset the good becomes? We address these questions by looking at the ways in which conversion from public good to public asset is achieved while still maintaining the characteristics of education as a public good provided through its non-excludable and non-rivalrous characteristics. The aim of what follows next is not to provide a detailed history of the relationship between the state and universities in England, but rather to signpost the changing logic of this relationship in terms of its evolution from how higher education became a public good to changes in higher education that enabled its gradual move toward being accounted for as an asset.

English Higher Education—Becoming a Public Good

The record of state involvement in higher education in England could be said to begin in the nineteenth century. Prior to this period of time, the two major centers—Oxford and Cambridge—were independent wealthy entities reliant on endowments with no financial support from the state. The nineteenth century saw the state's involvement in shaping English higher education through its legal interventions in Oxbridge which aimed to reorient these institutions toward national education, as part of a government move to expand university provision (Vernon 2004).

The direct and centrally coordinated financial support of higher education institutions through block grants—the initial transformation of university education into a public good—began with the establishment of the University Grants Committee (UGC) in 1919. The UGC was accountable to the Treasury until 1963 and then moved under the jurisdiction of the Department of Education and Science.[2] Until 1946, the UGC's role was to guide the allocation of "deficiency grants" designed to help financially when required (Shattock and Berdahl 1984, 472). However, in 1946 its purpose was reassessed with the view that "the time had come when the Government was bound to assure itself that somewhere in the University system provision is made for every field of scholarship or science which is necessary to the *national interest*" (Hetherington 1954, cited in Owen 1980, 264, emphasis added). The Education Act of 1962 made higher education free by requiring local authorities (but effectively the Treasury) to pay university fees for students,[3] while the Robbins Report of 1963 introduced the idea of higher education as a citizens' right, thus articulating the concept of higher education as a public good. The government financing of higher education grew from 33.6 percent of all income received by universities (including endowments and fees) in 1921 to 76.4 percent in 1973 (Owen 1985, 46–47).

The steady increase in public financing of universities became problematic with the expansion of higher education. There were no new universities established between 1969 and 1992 in the UK, although higher education participation was growing steadily (Collini 2012).[4] This changed with government reform through the 1988 Education Reform Act and the 1992 Further and Higher Education Act. The reform reclassified polytechnics into universities and triggered the expansion of these new universities; as a result, the higher education participation rate in England rose from 15 percent in 1988 to 47 percent in 2014 (McGettigan 2013; UK Government 2015b). New legislation has also reformed the funding bodies: the University Grants Committee was replaced by the Universities Funding Council in 1989, and in 1992 the Higher Education Funding Council for England (HEFCE) was formed to oversee funding of English universities.[5]

The post-1992 rise in student numbers and the expansion of universities was a tipping point that led to a rethinking of higher education as a public good funded by tax revenues. Yet doubts regarding the sustainability of such funding had already set in during the late 1970s and early 1980s, as "the virtual zero growth of GNP [gross national product] has meant that increased education expenditure in real terms can only come at the expense of real reductions elsewhere" (Craven et al 1983, 579). Along with growing funding in the 1960s and 1970s came risk: exposure to government cuts resulting from economic fluctuations was the corollary of universities' dependence on public money (Anderson 2016).

The Introduction and Reclassification of Student Loans

Growing concerns over the sustainability of public financing of university teaching prompted the first introduction of student loans in 1990. Kenneth Baker (1986, cited in Wilson 1997, 12), at the time the Secretary of State for Education, reasoned that "student numbers in higher education are at an all-time record level [and] we want still more to benefit.... But in doing so we must have regard to the claims on national resources. That is why I think that the time is ripe to investigate with an open mind all possible forms and sources of support." Limited national resources became a basis for justifying the introduction of alternative sources of support for students—in the form of loans. However, these changes in the government's vision of how higher education ought to be financed did not start with the introduction of tuition fees payable by students to cover the costs of courses (that would come later). Making students pay expenses previously settled by government was initially implemented through the introduction of loans to cover students' costs of living (rather than the cost of courses) and these loans were designed to make up the difference that resulted from a freeze in the annual increase of student maintenance grants (Barr 1989).[6]

Through the Education (Student Loans) Act of 1990 (UK Government 1990) maintenance loans were introduced. These took the form of fixed-term mortgage-style loans with sixty equal monthly payments that had to be made once the borrower's income was over 85 percent of national average earnings. Once this threshold was met, monthly repayments ensued, depending on the size of the outstanding balance rather than any subsequent fluctuations in the loanee's salary. The loans were implemented in September 1990 and the Student Loan Company (SLC) was established to handle the administration of these and all subsequent loans (Hillman 2013).

However, this solution to the rising cost of higher education was not viewed entirely as a success. Among other concerns with higher education spending, was the so-called classification problem (Barr and Crawford 1998) of how to list (or classify) loans on government account books. Inquiring into how higher education funding should be organized in the UK, the Dearing Report (1997) made a number of recommendations. The report is mainly known in the UK for its recommendation to introduce a tuition fee to cover 25 percent of the cost of tuition. However, it also contained a seemingly obscure and technical argument. Recommendation 80 urged the government to consider adopting a different method to account for or classify student loans. The report advised: "Do not treat the repayable part of loans in the same way as grants to students." The maintenance loans had been classified as public spending on government accounts. This paid no recognition to the fact that some of these loans would be repaid in the future. The problem would seem to be urgent, but solvable: "The Report is spot-on in saying that this problem needs to be fixed, and fixed fast. If not resolved, it is terminal. It is true that loans will bring in additional resources from around 2020—but you cannot revive a corpse. Resolution, in contrast, will release a 'pot of gold' of over £1 billion, *immediately* and *every year*" (Barr and Crawford 1998, 75, emphasis in original).

This act of reclassification of student loans needed to be accommodated within the accounting framework used by the UK government. At the time of the Dearing Report, UK government accounting reform was in full swing.

It started in 1993 in order to account for the complex nature of government transactions. Founded in 1866, government accounting practice had been calculating profits and losses on a cash basis, focusing on cash flows in real time. In this way, money was accounted for when it was received or paid out, whereas "accruals accounts record costs and revenues as they are respectively incurred and earned. By contrast, cash accounting records cash payments and receipts when they are made" (Likierman 1995, 563).

Moving from cash accounting to the private sector technique of accruals accounting meant that the books showed government performance during a financial year, rather than merely recording cash flows. Most importantly, it provided a tool to account for student loans by classifying them as "financial transactions" instead of an outright expenditure (which was how the student grants had been accounted for). From the financial year of 2001–02 the Department for Education and Skills (the Treasury's responsibility) was fully reliant on resource (accruals) accounting (RA) (Heald 2005).

The End of Direct Funding of English Undergraduate Teaching and the Income Contingent Repayment (ICR) Loans

Facilitated by the UK government accounting reform that enabled the reclassification of student loans, the transformation of English higher education from a public good continued with what Hillman (2013, 259) called "the triumph of tuition loans." But these triumphant loans introduced in 1998 had a specific nature that differed from their predecessors. Whereas loans launched in 1990 were fixed-term mortgage-style loans, from 1998 they were *income-contingent repayment (ICR) loans* covering living costs only. As with mortgage style loans, repayments would start only once a graduate's income was over a certain threshold (although this changed from 85 percent of national average earnings to a threshold of minimum earnings of £10,000 a year rising to £15,000 in 2003). What was new was income contingency. This meant that monthly repayments would be linked to a graduate's salary—9 percent on total earnings—rather than the size of the outstanding debt. Moreover, outstanding loans would be written off after a number of years.

Further changes followed. To cover "deficiencies in the university estate [of] £11bn" (Dearing 2004), tuition fees rose to £3,000 a year in 2006. This reform caused controversy in parliament, where the Higher Education Bill passed its second reading by only 5 votes (UK Parliament 2004). But, equally important, up-front tuition fees that had been in place since 1998 were now to be covered by the ICR loans. Here is how Nicholas Barr, a key figure in the 2006 reform (Barr and Crawford 2003), describes the events:

> The introduction of small mortgage type loans in 1990 was a response to fiscal pressures from the growing system. The 1997 Dearing Report said, in effect: "Loans are the right way to go, but income contingent loans, not mortgage loans." This was Iain Crawford's and my great victory. Dearing had a rational strategy of income contingent loans, and fees of £1,000 covered by loans. That was a strategy—more cautious than I wanted, but a genuine stepping-stone. The government then subverted the strategy by introducing fees but without loans to cover them. The next round of reform was 2006, which was the one time that the government stuck to its strategy, because we had an education minister, Charles Clark, who was bright enough to understand the idea of a strategy and sufficiently a political big beast to be able to protect it from cherry-picking. So, the 2006 reforms included income contingent loans to cover living costs, variable fees of up to £3,000 fully covered by income contingent loans, and pro-access policies earlier in the system. That was a proper strategy. (Barr interview)

The strategy of ICR loans was designed to maintain higher education as a public good: resolving *problems of access* and *problems of fairness*. The problem of access posed the question of how to widen higher participation rates, given that the cost of a university degree is high and the return is not guaranteed:

> Suppose that you borrow to buy a house and the repayments are £500 a month. If your income falls you can sell the house and repay the loan. If you borrow to finance a degree and the repayments are £500 a month, the risk you take is that if your income falls, the repayments will be an intolerable burden. So what you are going to do? You are not going to borrow. Or you are not going to borrow enough. And this is Milton Friedman writing in 1955. . . . The income contingent formula protects the borrower against low income this month, forgiveness after thirty years protects him/her from low life-time income. Thus the argument is the risk you face is contained. (Barr interview)

Nicholas Barr refers to the work of Milton Friedman (1955) here as a means to introduce the economic logic of ICR loans: that the policy points to a way of solving a market imperfection that results in underinvestment in human capital. The source of the imperfection is tied to mortgage-style "fixed money loans" that have to work without collateral—an asset that could be retained to reimburse the lender if the borrower cannot continue with repayments. The absence of collateral means the loans are risky to the (private) lender who might compensate the risk by high interest, but that would deter the borrower. The mortgage style terms of the loan are also risky to the borrower, discouraged by fixed payments regardless of her future income. The lender could be encouraged to invest by "buy[ing] a share in an individual's earning prospects," but administering such investment is very costly in terms of monitoring the location and income of borrowers as well as the long term of the loans.[7] The resulting market imperfection gives grounds for government to become the issuer of income contingent loans.

The second problem that ICR loans are designed to solve is the problem of fairness concisely captured by former Conservative Minister for Universities and Science David Willetts (2015, 14): "Even though there are public benefits from a graduate going into a very-well-paid job, it is not clear that on its own it justifies less affluent tax-payers subsidising it. Repayments by graduates who enjoy earnings above the average as a result of their university education appears fair—otherwise lower income non-graduate tax payers would be meeting the cost of a university education." Solving these two problems—of access and of fairness—through ICR loans resulted in the state covering only for a proportion of the costs of higher education provision, in particular when loans are not repaid. This public subsidy has continued and enabled English higher education to (at least nominally) retain its status as a public good.

However, the financial crisis of 2008 and the emergence of austerity measures in UK politics have led to further change. The Browne Report (2010) recommended a course of action to optimize the financing of the provision of higher education in England. As a result the UK coalition government of the Conservatives and Liberal Democrats initiated a reform in how it financed higher education in England. If the tuition fees introduced in 1998 and raised in 2006 had been providing resources in addition to what English universities were receiving (in the form of the direct block grant for teaching from the government), the increase to £9,000 a year in tuition fees for full-time UK and EU students was set to replace direct government funding in "low-cost" subjects and a significant reduction of government funding in "high-cost" subjects.[8]

The existing income contingent repayment (ICR) loans were modified to fit the purpose of the reform. The salary threshold at which repayments would kick in was raised to £21,000, and the write-off period was extended to thirty years (Cartwright 2016). Interest paid on outstanding loans taken after September 2012 varied between the Retail Price Index (RPI, when income was below the threshold) and RPI plus 3 percent (while students were studying, but also once income is £41,000 or more, rising to 4.6 percent in 2016). For pre-2012 loans the interest rate was not changed (SLC 2016).[9] Andrew McGettigan, an expert on UK higher education and its financing, explains:

> There are good arguments about why you do have a real interest rate on student loans, because then you are not subsidising wealthy people. Because the only people who ever likely to pay any of the interest rather than principal on these student loans are very high earners. So if you have zero interest rate, you are basically letting very high earners get out of the system quicker.... Most people in finance get this wrong—they think that the cost of student loans is the interest, and yet this cannot be, because the income contingent structure [of the loans]. The reason why the interest rate—while people are studying—is RPI plus 3 percent is to stop people from wealthy background having a subsidised loans for three years while they are studying, invest it somewhere else, then pay it off as soon as they graduate. (McGettigan interview)

Since 2012–13, the size of the issued income contingent repayment loans has surged, prompted by the near tripling of tuition fees (from just over £3,000 to £9,000), mostly covered by the ICR loans, and lifting of the cap on the number of high-performing students a university can enroll. In 2015–16 alone the amount lent to students was £11.8bn (rising from £6bn in 2011–12), with total outstanding balance being £76.3bn (compared with £39.6bn in 2011–12; Cartwright 2016).

Income Contingent Repayment (ICR) Loans as an Impaired Public Asset

Given the scale of lending, and the fact that the loans have been reclassified on the government books as a financial asset that produces future revenues rather than a direct and irredeemable cost, the question of crucial importance is how to *value* repayments to government made in thirty years? When a sum of money is lent in a conventional way, fixed-period repayments start coming in immediately after borrowing and, given the interest paid on the borrowed amount, bigger repayments are made by the borrower in the early years of the loan. This certainty makes it quite straightforward to know the full sum borrowed together with interest and when the outstanding balance will be cleared. However, the income contingent nature of student loan repayments has a corollary. An ICR loan is a long-term (thirty-year) asset, with most of the repayments accruing in the future (once, and if, the graduate's income rises). Consider, for instance, an ICR loan of £40,351 a UK student took out in 2012 to cover their tuition fees and maintenance expenses for three years of study. The highest growth of their real earnings and thus highest repayments are estimated to be coming in 2027–29 (Shephard 2013, 3), while a certain proportion of the loans will not be repaid in full, or at all, if the graduate's income is never above the threshold. In other words, the financial asset will not generate value equal to what is spent on issuing the loans today or, using economic terms, the face value of the asset is not the same as its fair value, therefore the asset is impaired.

Since the impairment is the value that is never going to be returned to government, it is viewed as a cost (i.e., public spending on university teaching in England). As such it has various elements to cover for, which reflect not only nonpayments per se (due to death, disability, and income below the threshold) but also the cost of money used to issue the ICR loans (the interest rate subsidy). As a cost, the impairment needs to be budgeted for today. Yet how should this sum be calculated of what will not be repaid in the future? Keeping in mind that student loans are the biggest financial asset on the government books—it amounted to £76bn in 2016 (UK Government 2016)—it makes the techniques of valuation that will account for ICR loans, specifically for the size of the asset's impairment, of particular importance: "The accounting has changed several times with student loans. It's one of the most dynamic aspects of accounting.... Government is experimenting here; it's doing something that has not really been done before [because] there is no commercial expertise to draw on, there is no preexistent databases, so you need a good accounting system" (McGettingan interview).

A discount rate is a measure instrumental in resource accounting (the initial valuation practice that made it possible to consider student loans as an asset). A discount rate can accommodate time difference in the ICR loans' costs and revenues, and can thus be used to determine the size of impairment of ICR student loans. The impairment is called the Resource Accounting and Budgeting (RAB) charge by the UK government. To calculate this charge (or to put a figure on the government commitment to cover for what will not be returned by graduates) is a twofold task. Since the impairment is the difference in two values (face and fair), it should be found by subtraction that is only possible once a fair value, or net present value (NPV), of the future incoming payments is calculated (UK Government 2016a). For instance, if a graduate who took a loan in 2017 is forecast to pay £1,000 in 2027, how much is this worth now? This valuation is done by discounting payment forecasts back to the time of issuing the loans. This involves applying the discount rate produced by the Treasury department with the advice of the Financial Reporting Advisory Board (FRAB). The current discount rate that is applied to repayment forecasts is the Retail Price Index (RPI) plus 0.7 percent. So the graduate's payment of £1,000 in 2027 is worth £933 today:

$$\pounds 933 = \frac{\pounds 1000}{(1+0.007)^{10[\text{years}]}}$$

But this calculation is for illustration only. The precise monetary value is indicated "as a proportion of the initial loan outlay" and currently is between 41 percent and 52 percent (with an RPI higher than 1), which could be viewed as for each £1 lent the return is 48–59 pence (UK Government 2019). Since its first introduction as part of the funding policy in 2011, the RAB charge proved to be rather volatile. Initially estimated at between 28 percent and 30 percent, by the beginning of 2014 it had risen to 45 percent. Gavan Conlon, a partner in London-based economic policy consultancy that undertook RAB charge modeling, reflects on the change:

> Initially, the Browne [Review] suggested that the RAB charge would be unchanged, which is absolutely nonsensical. Then, over the next three to four years it was revised from 30 percent to 32 percent to 35 percent to 40 percent and it finished at 45 percent. The reason why the RAB charge estimate increased to approximately 45 percent was because it is vastly dependent on both the size of the loans offered to students (which increased sharply) but also graduate earning growth (which was much lower than expected). (Conlon interview)

In 2014, the Office for Budget Responsibility (OBR, an independent examiner of the UK public finances) lowered its forecast of loan repayments due to the low "earnings growth for less well-paid graduates." Coupled with higher tuition fees that pushed up the size of the loans, the rise of the RAB charge caused controversy as it would seem that the government "has saved little or no public money by trebling fees to £9,000 and scrapping direct [teaching] grant" (Morgan 2014).

In March 2015, the Institute for Fiscal Studies (IFS) issued a report where it calculated the way the RAB charge could be lower if the discount rate was RPI+1.1 percent rather than RPI+2.2 percent (the discount rate used in RAB calculation at the time) arguing that "the assumption made about the discount rate is a key driver of this debate": "What is clear from our analysis is that the discount rate matters hugely when estimating the cost of higher education. In fact, it matters more than plausible changes to the rate of real earnings growth. Reducing the discount rate means valuing future repayments more highly; hence the estimated loan subsidy (and the RAB charge) falls" (Britton and Crawford 2015).

A change in assumptions regarding future repayments of the loans had been called for since 2012, when a letter in the *Financial Times* argued that if the discount rate were linked to the actual cost of government borrowing, it would be "sufficient to completely eliminate the predicted losses" (Leunig and Shephard 2012). Indeed, the discount rate that the government used in their RAB calculation was set in 2006 and bore no relation to its cost of borrowing:

> If you are a commercial operator, or a company, you set [the discount rate] first in relation to your cost of borrowing, and also expectation of inflation, and alternative investment opportunities, and you set your discount rate at your safest investment opportunity [but] this was never set in relation to cost of borrowing [and] we don't know which gilts [bonds issued by the government] have paid for student loans. (McGettigan interview)

In December 2015, the Treasury announced that "the real financial instrument discount rate to be applied at 31 March 2016 is 0.7 [and] the rate as applied to flows expressed in current prices is RPI+0.7, where the financial instrument is index linked to RPI" (UK Government 2015a). The reduction of the discount rate drove the RAB charge from 45 percent down to 20–25 percent and "by the stroke of the pen all those estimates, all the figures have become happy figures":

> It undercuts the Labour party, their whole strategy gets undermined, because they're attacking the RAB charge, suddenly the RAB charge drops 20 percent—nothing to attack. Strategically, politically it's brilliant and obvious. At the same time it changes the budgeting for the Department for Business and Innovation, and for the Department for Education now that got responsible for student loans—they now don't have a budget crisis. (McGettigan interview)

The "budget crisis" here relates to the Departmental Expenditure Limit (DEL) of the department responsible for HE (it used to be the Department for Business, Innovation and Skills, or BIS). Each year the Treasury allocates a certain amount of budget to cover the RAB charge for the new loans issued that year. At the same time the existing loans are revalued, and if their value is less than budgeted for then the department covers the difference from a "stock charge," which is part of the RAB charge within their DEL. However, since the RAB charge was growing quickly and steadily, and Department budgets until the end of 2015 were set in 2010, the BIS department needed more money to cover for it. As a result, the accounting rules changed in 2013 to provide additional funds for unexpected volatility—Annually Managed Expenditure (AME). This has to be "serviced." The Treasury charges BIS "a thirtieth of it every year for 30 years, it's like an internal loan but within your own budget," which needs to be paid with real money and "this payment goes in cash from their budget." The "lower discount rate ... reduce[d] the spread between the target impairment and the current RAB charge, which ... then translate[d] into a smaller spillover into other expenditure reductions" (McGettigan interview; McGettigan 2015a, 41).

Given the volatility in size of the impairment that is, arguably, arbitrarily managed through the discount rate, as well as the contingent nature of the rules governing budget that covers for the asset's impairment, the technique by which the ICR loans are accounted for is highly consequential. But, as Nick Hillman pointed out, "there [were] a number of reasons of why the RAB charge fell, it was not just the discount rate" (Hillman interview). He explained during a hearing in the House of Lords Select Committee on Economic Affairs: "The system has some flex built into it: you can change the terms and conditions of the loans" (House of Lords 2016). Subsequently Philippa Lloyd, the director general of higher and further education at the Department for Education (responsible for the loans since July 2016), was asked whether the question of higher education funding is gradually becoming a question of "managing the loan book." She replied, "You may decide to adjust levers in order to keep it [the higher education funding policy] on a sustainable footing" (House of Lords 2016). For example, one of the "levers" was the earning threshold of £21,000 beyond which graduates would have to start repaying loans. This was controversially frozen for five years (2015–2020) despite a promise to change it according to national average earnings (Elgot 2016). Managing the loan book thus involves steering higher education funding in certain directions and, as McGettigan (2015a, 43) argues, "is pushing policymakers towards certain solutions, which may not be in the general interest of universities and colleges or students."

Conclusion

In this chapter, we have explored the ways in which the reorganization of government funding for English higher education has created a public asset. In doing so, we have steered away from more general public and academic debates on English higher education focused on marketization (Molesworth et al. 2011), neoliberalization (Canaan and Shumar 2008), or financialization (Holmwood 2014). Instead, we oriented notions of financialization (Chiapello 2015; Engelen 2008; Van Der Zwan 2014; Davis and Kim 2015) toward problems tackled within the public sector and drew our inspiration from studies of the practices of economic valuation in public goods (Roscoe and Townley 2016; Doganova and Laurent 2016; Pallesen 2016; O'Brien 2016) and through forms of capitalization (Muniesa 2012) and assetization (Birch 2017). From here, we suggested an expansion of pragmatics of valuation by returning to Dewey (1937) and paying close attention to what is done to what in the accomplishment of assets. Although this has required a detailed study of the mundane details of the seemingly technical processes of assetization, we suggest this is immensely important given the long-lasting and highly political consequences of these processes that indeed give the new meaning to "the role and subjectivity… of public service users and providers" (Mennicken and Muniesa 2017).

Our empirical focus drew together complex successive actions through which English higher education and student loans came to occupy a particular asset form (see Birch and Muniesa, this volume). This focus on succession enables us to move away from any sense that these policies followed a single logic or intent. As Nick Hillman, the former chief of staff and special adviser for the Department for Business, Innovation and Skills once put it, "The fact that higher fees could make higher education more like a regulated market, with students coming to resemble consumers, was a bonus, but it was not the primary purpose" (Hillman 2016, 338–339). In this way, interventions in English higher education were successive in the sense that each basis for putting forward a means to resolve an issue built on previous attempts at resolution and on-going concerns raised regarding the form of resolution.

What began, then, as an articulated concern over the sustainability of public financing of university teaching came to be understood as resolvable through mortgage style fixed repayment loans. Yet the loans appeared to trigger a "classification problem" (Barr and Crawford 1998)—unlike other forms of government spending, these were not a straightforward cost but an initial outlay that would at some point bring in a return. The classification problem coincided with governmental accounting reform and the switch from cash accounting to accrual or resource accounting that took place over several years of UK government action. At the same time, concerns were raised regarding the fixed thresholds and fixed repayment terms of the loans, which seemed to undermine government policy and universities' hopes to ensure fair access to higher education. The public good was getting in the way of the financial asset.

Issuing Income Contingent Repayment (ICR) loans became a means to apparently repair this deficiency. With repayments indexed to fluctuations in salary and debts canceled after a fixed period, the public good would be transformed into an asset. Yet this also made it more challenging to value the emerging asset, as the size of repayments, length of term and even the possibility of any payments at all, were rendered less secure. Although various accounting techniques (such as discount rates) could be used to calculate a present value for these future repayment-based income streams, these values were below the amount lent to students. Accounting techniques made the loans appear as a significant cost to government, especially as student numbers and then loan amounts increased significantly in a short space of time, particularly when tuition fees were tripled. Although a RAB charge could be used to cover this shortfall, economists made clear in their discussion of the discount rate and the necessity to lower it: "note that nothing '*real*' has changed. No additional repayments are being made; we have simply changed how highly we *value* these future repayments in the present" (Britton and Crawford 2015, emphasis added).

Our close focus on the pragmatics of valuation draws attention to what is done to what in the constitution of this peculiar impaired asset. Yet its impairment appears to be an inescapable feature of the type of asset it became. Moreover, ultimately, the tenacious impairment would seem to become so looming that the process of turning the public good into an asset has been reversed. In December 2018, the UK Office for National Statistics ruled that the way income contingent student loans are accounted for, specifically the RAB charge, is to be changed and the loans are to be "reclassified as public spending" (Coughlan 2018). Student loans proved to be not just any asset, available to be made sense of through standard techniques of accounting. Impairment results from the continuing need of the UK government to cover the costs of loans' income contingency that provides the current UK government basis for dealing with fair access. In this way, impairment fundamentally affected the ability of higher education to move from a public good to an asset that retains (at least nominally) its goodness: a public asset.

Notes

1. We are aware of the fact that the concept of higher education as a (nonexcludable) public good could be criticized given that it is different from compulsory primary and secondary education as higher education is "a matter of choice" (Barr 2012, 491). However, here we term English higher education a public good because until 2012 for the most part it had been financed through general taxation.

2. The Treasury as well as the Board of Education had been providing some financial assistance to universities in the form of grants since 1907 (Owen 1980). Also, the UGC's influence on higher education in England and the UK went beyond direct allocation of financial support, but given the focus of the essay we discuss its financial activity.

3. Granted that the tuition fee payments by local authorities "were partially means-tested until the late 1970s" (Hillman 2013, 251).

4. With the exception of the University College of Buckingham, which was founded in 1976 as an independent (not financed by the state) higher education institution (Shattock 1994).

5. The 1992 Further and Higher Education (Scotland) Act made separate HE funding provisions for Scotland.

6. By 1989 with the introduction of the Research Assessment Exercises (RAEs had been initiated by the UGC in 1985) the financial provisions (block grants) for teaching and research were set apart and, following the change from the UGC to HEFCE, the system of financial provision for higher education that had been formed through the establishment of the UGC as a main body providing vision for as well as management of university funding, ceased to exist (Shattock 1994).

7. Notably, the unusual nature of the loans is said to be the reason the agency that currently administers UK ICR loans, the Student Loan Company—owned by the UK government and devolved administrations—is called a company. According to Willetts (2015, 18) the name was initially chosen in "the expectation that the clearing banks would co-own it and lend the funds [but] they backed out because the scheme was so different from their usual commercial lending."

8. The "low-cost" subjects are Arts and Humanities, Mathematics, Law, and Business, whereas "high-cost" subjects are Clinical Medicine, Science, Engineering, and Technology. McGettigan (2013, 27) shows how the funding for full-time undergraduate study per student fell from £3,898 and £2,709 for "low-cost" subjects in 2011–12 to no government funding in 2012–13, whereas the funding for "high-cost" subjects was significantly reduced—from £14,601 and £5,484 to £10,000 and £1,500, respectively.

9. Although there was a proposal to increase the interest rates for pre-2012 loans in order to make the loan book attractive for private purchasers in government asset sell-off (Chakrabortty 2013; McGettigan 2015b).

List of Interviewees

Nicholas Barr, November 25, 2016. Skype interview.

Gavan Conlon, July 28, 2016. London.

Nicholas Hillman, November 30, 2016. London.

Andrew McGettigan, December 8, 2016. London.

References

AAAS (American Academy of Arts and Sciences). 2016. *Public Research Universities: Serving the Public Good*, https://www.amacad.org/multimedia/pdfs/publications/researchpapersmonographs/PublicResearchUniv_PublicGood.pdf.

Anderson, R. 2016. University Fees in Historical Perspective. *History & Policy*. Policy Paper, February 8, 2016, http://www.historyandpolicy.org/policy-papers/papers/university-fees-in-historical-perspective.

Barr, N. 1989. The White Paper on Student Loans. *Journal of Social Policy* 18 (3): 409–17.

Barr, N. 2012. The Higher Education White Paper: The Good, the Bad, the Unspeakable—and the Next White Paper. *Social Policy & Administration* 46 (5): 483–508.

Barr, N., and Crawford, I. 1998. The Dearing Report and the Government's Response: A Critique. *Political Quarterly* 69 (1): 72–84.

Barr, N., and Crawford, I. 2003. Myth or Magic? *Guardian* December 2, 2003, https://www.theguardian.com/education/2003/dec/02/highereducation.tuitionfees.

BBC News. 2012. Senior Lib Dems Apologise over Tuition Fees Pledge, http://www.bbc.co.uk/news/uk-politics-19646731.

BBC News. 2016. Lib Dems Blame General Election Defeat on "Perfect Storm," http://www.bbc.co.uk/news/uk-politics-35615285.

Birch, K. 2017. Rethinking Value in the Bio-economy: Finance, Assetization, and the Management of Value. *Science, Technology, & Human Values* 42 (3): 460–490.

Birch, K., and Tyfield, D. 2013. Theorizing the Bioeconomy: Biovalue, Biocapital, Bioeconomics or…What? *Science, Technology, & Human Values* 38 (3): 299–327.

Britton, J., and Crawford, C. 2015. *There's More to Higher Education Funding Than the RAB Charge*. Institute for Fiscal Studies, https://www.ifs.org.uk/publications/7673.

Callon, M. 1994. Is Science a Public Good? *Science, Technology, & Human Values* 19 (4): 395–424.

Canaan, J., and Shumar, W., eds. 2008. *Structure and Agency in the Neoliberal University*. Abingdon: Taylor & Francis.

Cartwright, D. 2016. *Student Loans in England: Financial Year 2015–2016. Statistical First Release*. Student Loans Company (SLC), June 16, 2016, http://www.slc.co.uk/media/7594/slcsfr012016.pdf.

Chakrabortty, A. 2013. Raise Interest Rates on Old Student Loans, Secret Report Proposes. *Guardian*, June 13, 2013, https://www.theguardian.com/money/2013/jun/13/raise-interest-rate-student-loans-secret-report.

Chiapello, E. 2015. Financialisation of Valuation. *Human Studies* 38 (1): 13–35.

Collini, S. 2012. *What Are Universities For?* London: Penguin Books.

Coughlan, S. 2018. Student Loan Ruling Adds £12bn to Government Borrowing. *BBC News*, December 17, 2018, https://www.bbc.co.uk/news/education-46591500.

Craven, B., Dick, B., and Wood, B. 1983. Resource Allocation in Higher Education in Britain. *Higher Education* 12 (5): 579–589.

Davis, G., and Kim, S. 2015. Financialization of the Economy. *Annual Review of Sociology* 41: 203–221.

Dearing, R. 1997. *Higher Education in the Learning Society*. London: Her Majesty's Stationery Office 1997, http://www.educationengland.org.uk/documents/dearing1997/dearing1997.html.

Dearing, R. 2004. We Will All Benefit from Tuition Fees. *Guardian*, January 8, 2004, https://www.theguardian.com/politics/2004/jan/08/publicservices.studentpolitics.

Dewey, J. 1937. *Logic—The Theory of Inquiry*. Carbondale: Southern Illinois University Press.

Doganova, L., and Laurent, B. 2016. Keeping Things Different: Coexistence within European Markets for Cleantech and Biofuels. *Journal of Cultural Economy* 9 (2): 141–156.

Doganova, L., and Muniesa, F. 2015. Capitalization Devices: Business Models and the Renewal of Markets. In *Making Things Valuable*, edited by M. Kornberger, L. Justesen, A. Koed Madsen, and J. Mouritsen, 109–125. Oxford: Oxford University Press.

Elgot, J. 2016. Student Loans: Fight over "Disgraceful" Freeze on Income Threshold. *Guardian*, November 22, 2016, https://www.theguardian.com/money/2016/nov/22/student-loans-fight-over-disgraceful-freeze-on-income-threshold.

Engelen, E. 2008. The Case for Financialization. *Competition & Change* 12 (2): 111–119.

Friedman, M. 1955. The Role of Government in Education. In *Economics and the Public Interest*, edited by R. Solo. New Brunswick, NJ: Rutgers University Press.

Gil, N. 2015. LSE Takes Legal Action to Evict Occupying Student Protesters. *Guardian*, April 30, 2015, https://www.theguardian.com/world/2015/apr/30/lse-takes-legal-action-to-evict-occupying-student-protesters.

Hatherly, D., Leung, D., and MacKenzie, D. 2008. The Finitist Accountant. In *Living in a Material World,* edited by T. Pinch and R. Swedberg, 131–160. Cambridge, MA: MIT Press.

Heald, D. 2005. The Implementation of Resource Accounting in UK Central Government. *Financial Accountability & Management* 21 (2): 163–189.

Helgesson, C.-F., Lee, F., and Lindén, L. 2016. Valuations of Experimental Designs in Proteomic Biomarker Experiments and Traditional randomized Controlled Trials. *Journal of Cultural Economy* 9 (2): 157–172.

Helgesson, C.-F., and Muniesa, F. 2013. For What It's Worth: An Introduction to Valuation Studies. *Valuation Studies* 1 (1): 1–10.

Hillman, N. 2013. From Grants for All to Loans for All: Undergraduate Finance from the Implementation of the Anderson Report (1962) to the Implementation of the Browne Report (2012). *Contemporary British History* 27 (3): 249–270.

Hillman, N. 2016. The Coalition's Higher Education Reforms in England. *Oxford Review of Education* 42 (3): 330–345.

Holmwood, J. 2014. Beyond Capital? The Challenge for Sociology in Britain. *British Journal of Sociology* 65 (4): 607–618.

House of Lords. 2016. *Inquiry on Student Loans*: *Revised transcript of evidence taken before the Selected Committee on Economic Affairs,* July, http://data.parliament.uk/writtenevidence/committeeevidence.svc/evidencedocument/economic-affairs-committee/student-loans/oral/35022.html.

IASB (International Accounting Standards Board). 2015. *International Financial Reporting Standard for Small and Medium-Sized Entities,* http://eifrs.ifrs.org/eifrs/sme/en/IFRS%20for%20SMEs_Standard_2015.pdf.

Kornberger, M., Justesen, L., Koed Madsen, A., and Mouritsen, J., eds. 2015. *Making Things Valuable.* Oxford: Oxford University Press.

Latour, B. 1987. *Science in Action.* Cambridge, MA: Harvard University Press.

Leunig, T., and Shephard, N. 2012. The Cost of Expanding Higher Education Is—Zero. *Financial Times,* February 8, 2012.

Lewis, P., Vasagar, J., Williams, R., and Taylor, M. 2010. Student Protest over Fees Turns Violent. *Guardian,* November 10, 2010, https://www.theguardian.com/education/2010/nov/10/student-protest-fees-violent.

Leys, C. 2001. *Market-Driven Politics: Neoliberal Democracy and the Public Interest.* London: Verso.

Likierman, A. 1995. Resource Accounting and Budgeting: Rationale and Background. *Public Administration* 73 (4): 562–570.

McGettigan, A. 2013. *The Great University Gamble: Money, Markets and the Future of Higher Education*. London: Pluto Press.

McGettigan, A. 2015a. *The Accounting and Budgeting of Student Loans*. Higher Education Policy Institute. Report 75.

McGettigan, A. 2015b. Cash Today. *London Review of Books*, March 5, 2015, http://www.lrb.co.uk/v37/n05/andrew-mcgettigan/cash-today.

Mennicken, A., and Millo, Y. 2016. *Testing Asset Values: Financialization, Organization and the Emergence of Asset Impairment Rules*. Presentation at 4S/EASST Conference: Science and Technology by Other Means. September, Barcelona, Spain.

Mennicken, A., and Muniesa, F. 2017. Governing Through Value: Public Service and the Asset Rationale. *CARR: Risk & Regulation* 34.

Mirowski, P. 2011. *Science-Mart: Privatizing American Science*. Cambridge, MA: Harvard University Press.

Molesworth, M., Scullion, R., and Nixon, E., eds. 2011. *The Marketisation of Higher Education and the Student as Consumer*. London: Routledge.

Morgan, J. 2014. "Massive" Budget Hole Predicted as RAB Charge Rises. *Times Higher Education*, March 21, 2014, https://www.timeshighereducation.com/massive-budget-hole-predicted-as-rab-charge-rises/2012189.article.

Mountford-Zimdars, A., Jones, S., Sullivan, A., and Heath, A. 2013. Framing Higher Education: Questions and Responses in the British Social Attitudes Survey, 1983–2010. *British Journal of Sociology of Education* 34 (5-6): 792–811.

Muniesa, F. 2012. A Flank Movement in the Understanding of Valuation. *Sociological Review* 59 (2): 24–38.

Ng, I., and Forbes, J. 2009. Education as Service: The Understanding of University Experience Through the Service Logic. *Journal of Marketing for Higher Education* 19 (1): 38–64.

Nitzan, J., and Bichler, S. 2009. *Capital as Power: A Study of Order and Creorder*. London: Routledge.

O'Brien, D. 2016. What Price Evidence? The Ethics of Office and the Ethics of Social Science in British Cultural Policy. *Journal of Cultural Economy* 9 (2): 127–140.

OECD (Organisation for Economic Co-operation and Development). 2015. *Education at a Glance 2015*, http://www.oecd-ilibrary.org/education/education-at-a-glance-2015_eag-2015-en.

Ortiz, H. 2013. Financial Value: Economic, Moral, Political, Global. *HAU: Journal of Ethnographic Theory* 3 (1): 64–79.

Owen, T. 1980. The University Grants Committee. *Oxford Review of Education* 6 (3): 255–278.

Owen, T. 1985. Financing University Education in Britain. *Western European Education* 17 (3): 45–60.

Pallesen, T. 2016. Valuation Struggles over Pricing—Determining the Worth of Wind Power. *Journal of Cultural Economy* 9 (6): 527–540.

Phipps, C. 2014. Demonstrators in "Scuffles" with Police as Students March Against Fees—As It Happened. *Guardian*, November 19, 2014, https://www.theguardian.com/education/live/2014/nov/19/students-march-tuition-fees-live.

Quinn, S. 2017. "The Miracles of Bookkeeping": How Budget Politics Link Fiscal Policies and Financial Markets. *American Journal of Sociology* 123 (1): 48–85.

Roscoe, P., and Townley, B. 2016. Unsettling Issues: Valuing Public Goods and the Production of Matters of Concern. *Journal of Cultural Economy* 9 (2): 121–126.

Rose, N., and Miller, P. 1992. Political Power beyond the State: Problematics of Government. *British Journal of Sociology* 43 (2): 173–205.

Samuelson, P. 1954. The Pure Theory of Public Expenditure. *Review of Economics and Statistics* 36 (4): 387–389.

Samuelson, P. 1955. A Diagrammatic Exposition of the Theory of Public Expenditure. *Review of Economics and Statistics* 37 (4): 350–356.

Shattock, M. 1994. *The UGC and the Management of British Universities*. SRHE & Open University Press.

Shattock, M., and Berdahl, R. 1984. The British University Grants Committee 1919–83: Changing Relationships with Government and the Universities. *Higher Education* 13 (5): 471–499.

Shephard, N. 2013. *The Actual Financing Costs of English Higher Education Student Loans*, https://www.nuffield.ox.ac.uk/economics/papers/2013/FundingCosts20130508.pdf.

SLC (Student Loan Company). 2016. *Interest Rates*, http://www.slc.co.uk/services/interest-rates.aspx.

Stark, D. 2011. *The Sense of Dissonance: Accounts of Worth in Economic Life*. Princeton, NJ: Princeton University Press.

UK Government. 1990. *Education (Student Loans) Act*, http://www.legislation.gov.uk/ukpga/1990/6/contents/enacted.

UK Government. 2015a. *HM Treasury. Discount Rates for Post-Employment Benefits, General Provisions and Financial Instruments: Announcement of Rates*, https://www.gov.uk/government/uploads/system/uploads/attachment_data/file/488228/PES__2015_08_-_discount_rates_-_for_publishing.pdf.

UK Government. 2015b. *Participation Rates in Higher Education: Academic Years 2006/2007–2014/2015*. Department for Education, https://www.gov.uk/government

/uploads/system/uploads/attachment_data/file/552886/HEIPR_PUBLICATION_2014-15.pdf.

UK Government. 2016a. *Department for Business, Innovation & Skills: Annual Report and Accounts 2015–16*, https://www.gov.uk/government/uploads/system/uploads/attachment_data/file/537425/bis-annual-report-accounts-2015-16-web.pdf.

UK Government. 2016b. *Department for Business, Innovation and Skills: The RAB Charge*, https://www.gov.uk/government/uploads/system/uploads/attachment_data/file/501419/BIS-16-131-RAB-charge.pdf.

UK Government. 2019. Department for Education: Student Loans Forecasts, England, 2018 to 2019, https://assets.publishing.service.gov.uk/government/uploads/system/uploads/attachment_data/file/811997/Student_loan_forecasts_2018-19_-_text.pdf.

UK Parliament. 2004. *House of Commons Votes and Proceedings Tuesday 27th January 2004*, https://www.publications.parliament.uk/pa/cm/cmvote/40127v01.htm.

Van Der Zwan, N. 2014. Making Sense of Financialization. *Socio-Economic Review* 12: 99–129.

Vernon, K. 2004. *Universities and the State in England, 1850–1939*. Abingdon, Oxon: Routledge.

Weinstein, M. 2009. Paul A. Samuelson, Economist, Dies at 94. *New York Times*, December 13, 2009, http://www.nytimes.com/2009/12/14/business/economy/14samuelson.html.

Willetts, D. 2015. *Issues and Ideas on Higher Education: Who Benefits? Who Pays?* London: Policy Institute at King's College London.

Wilson, W. 1997. *Student Grants, Loans and Tuition Fees*. Research Paper No 97/119. House of Commons Library, http://researchbriefings.files.parliament.uk/documents/RP97–119/RP97–119.pdf.

11 Recidivists, Rough Sleepers, and the Unemployed as Financial Assets: Social Impact Bonds and the Creation of New Markets in Social Services

James W. Williams

Introduction

One of the enduring ironies of the financial crisis of 2008 is that these events did little to challenge neoliberal rationalities, with markets continuing to be viewed as a response to all manner of social problems (Mirowski 2013). A perfect expression of this valorization of market-based solutions comes from the social services sector and the emergence of a new, investment-based funding model, the social impact bond (SIB).[1] Pioneered in the UK in 2010, a SIB is an investment contract in which private investors provide up-front funding for a preventative program. If the program is successful in meeting predefined performance targets, the government repays the investment and provides a return based on the cost savings realized from reduced future demand on public services. The result is a win-win-win scenario. Nonprofits receive multiyear, more flexible funding. Governments are able to deliver preventative services while only paying for success. And investors receive a financial return while also doing good. These benefits have helped fuel the global expansion of the model, which has been exported to the US, Australia, Canada, and parts of Europe, and applied to a range of issues, including reoffending, homelessness, child welfare, education, and employment (Carter et al. 2018).[2]

As the SIB market has grown, so too has the number of critics. Commentators in policy and academic circles have cited concerns ranging from the failure to deliver on promised cost savings, to perverse incentives, to the erosion of agency autonomy and mission (Cooper et al. 2016; Edmiston and Nicholls 2018; Fraser et al. 2018a; McHugh et al. 2013; Neyland 2018; Warner 2013). A key theme running through many critical accounts is that SIBs are indicative of the marketization (Joy and Shields 2013; Sinclair et al.

2014) or, more specifically, the financialization of the nonprofit and public services sector (Cooper et al. 2016; Lake 2016; Warner 2013), a migration of financial actors, tools, and logics into the nonprofit space and the transformation of human services and their clients into commodities (Cooper et al. 2016) and financial assets (Dowling 2017; Dowling and Harvie 2014). And yet, while SIBs are clearly informed by a financial logic and are enlivened by the interests of prospective investors, the notion that this model is yet another example of the expanding remit of finance overlooks the unique features of this space and the hard work required to reconfigure social services as investment propositions. Far from a smooth process of financialization, SIB development has encountered a series of roadblocks with practitioners at times struggling to engage investors and the market growing much more slowly than expected. All of this leads to a slightly different question. Rather than "What is wrong with SIBs?" or "Are SIBs good or bad?" we might ask, "Why have SIBs struggled to take hold?"

Informed by the results of a larger three-year study of SIBs and the funding of nonprofit social services in Canada, the US, and the UK, this chapter provides a tentative answer to this question by framing SIB development not in terms of a process of financialization but rather assetization. Drawing from the core themes of this collection, what is most noteworthy about SIBs is not simply the transformation of social problems into investment propositions but rather a distinct form and practice of valuation through which the work of nonprofits is reconfigured as a type of asset yielding savings to government and returns to investors. In undertaking this work, SIB specialists have faced a variety of challenges struggling with how to value these transactions and to build a value proposition sufficient to engage investors, government, and providers both in individual deals and in a longer-term commitment to the market. The central argument of the chapter is that it is these valuation challenges which help to account for the struggles of the SIB market to grow and gain traction, with assetization thus providing an invaluable lens through which to explore the contingencies and limits of the SIB phenomenon. The first part of the chapter lays the groundwork for this analysis, providing an overview of the SIB model and the valuation practices that have been critical to turning social services into assets. The second part then turns to the struggles underlying this process of assetization, examining three distinct valuation challenges that have impacted SIB development. Given these challenges, and in light of the recent evolution of the SIB market, the chapter concludes with the suggestion that what may be most significant about SIBs moving forward is the creation not of a *private asset* but rather a new type of *public asset*, one that is likely to have critical implications for the relationship between the state and the social sector (see Milyaeva and Neyland, this volume).

Financing Social Services: SIBs as a New Funding Tool

In September 2010, the UK government introduced with great fanfare the pilot of a new model for funding social services. Termed a "social impact bond," the concept was to use private capital to fund an initiative designed to reduce reoffending among short-term prisoners released from Peterborough prison. If the program was successful and met predetermined performance targets (defined in terms of reductions of recidivism rates relative to a matched comparison group), investors would receive a financial return of up to 13 percent (Disley et al. 2011). This return would be paid by the government based on the cost savings realized from reduced future demand on the criminal justice system. However, if the program failed to meet these targets, investors would lose their principal along with any potential returns. This transaction was designed by an intermediary organization, Social Finance UK, established with the express purpose of building the SIB market, and the outcomes were measured and validated by an independent evaluator.

Often described as a game changer in the funding of social services, this model is seen to provide several benefits. For nonprofits, it offers a source of flexible, longer-term funding that is superior to standard government contracts. For government, it provides a way to deliver more innovative and preventative programs while shifting the up-front costs and, more importantly, the risk onto private investors. And, for investors, SIBs offer the promise of both a financial and a social return, exactly the type of blended value proposition heralded within the emerging fields of "social finance" and "impact investing." The ultimate vision is that SIBs will help unlock new sources of investment capital for the purpose of social good. Based on the appeal of this win-win-win scenario, the model has quickly expanded to the US, Canada, Australia, and parts of Europe, with the SIB market surpassing one hundred operational projects worldwide as of the midpoint of 2018 (Carter et al. 2018).

As the slate of projects has grown, so too has the chorus of critics, with SIB detractors citing a host of flaws and concerns ranging from perverse incentives to the challenge of evaluating social programs, to high transaction costs, to threats to the autonomy, mission, and moral mandate of providers (Cooper et al. 2016; Fox and Albertson 2011; Joy and Shields 2013; McHugh et al. 2013; Neyland 2018; Warner 2013). A key thematic current underlying many of these critical accounts is the notion that SIBs are indicative of the financialization of social and public services (Dowling 2017; Lake 2016; Warner 2013), a migration of financial actors, interests, and logics into the nonprofit space with social services thus transformed from public goods into a "new asset class" and form of capital accumulation (Dowling 2016, 6; Dowling and Harvie 2014; Lake 2016).

There is without question merit to these critiques. It is undeniable that SIBs are enlivened by, and have their roots in, the logics of finance. By their very design, they provide a mechanism through which investors are able to profit from social problems and, with projected returns in the 3 percent to 15 percent range, the potential for profit is very real. And yet, the reality of SIBs as they have evolved in the UK, US, and Canada bears only a limited resemblance to the financialization narrative. The anticipated influx of investors has largely failed to materialize, with more return-motivated investors remaining quite cool to the model (Fraser et al. 2018b). Instead, SIB markets have relied heavily on state and philanthropic support and subsidy (Floyd 2017; Warner 2013). In the UK, the central government has been a key backer of SIBs supplying investment capital through the social investment wholesaler Big Society Capital. In the US, philanthropy has played an essential supporting role, with major foundations providing direct investments, loan guarantees,[3] and first loss capital.[4] In addition to the struggle to engage more commercial capital, there are other signs of weakness in the SIB market. While the number of transactions continues to grow, the pace of growth has been slower than expected (Maier and Meyer 2017; Arena et al. 2016) and the market as a whole remains tiny, especially when compared to the scale of government spending (Dey and Gibbon 2018; Giacomantonio 2017). Many projects have also failed to launch, foundering between feasibility and execution (Fraser et al. 2018b; Heinrich and Kabourek 2018), and several advisory firms have struggled with lower than expected revenues and uncertainty around the sustainability of their business models.[5]

The fact that SIBs have struggled to live up to expectations, showing few signs of the kind of capital inflows and market growth imagined by advocates and feared by critics, suggests that perhaps what needs to be explained is not the existence of SIBs as a vehicle for the financialization of social services, but rather the challenges, barriers, and limits associated with the attempt to bring an investment model and mindset to bear on the nonprofit sector. While often overlooked in the academic literature, the trials and tribulations of SIB development and execution are key themes in practitioner reports (e.g., Dear et al. 2016). Beyond common complaints about a lack of capacity and expertise on the part of government and providers as well as difficulties around data access, central to many of these accounts is the notion that SIBs are a great deal of work and that much of this work involves the effort to capture and quantify the economic and financial value of nonprofits, thus allowing for program outcomes to be monetized and capitalized as savings for government and returns for investors. This suggests that what lies at heart of SIBs is a distinct challenge of valuation, or value creation, with SIB specialists employing a range of different tools and forms of expertise (e.g., accounting, cost-benefit analysis, program evaluation, data analytics, and performance management), few of which are strictly financial in nature. The centrality of valuation to the SIB enterprise, and the nature of SIBs as a claim on the projected future value of nonprofit work, suggest that SIBs may be more usefully examined through the lens of assetization rather than financialization.

As described by Birch and Muniesa (this volume), the process of turning things into assets depends on particular practices of valuation. This follows the core insight of valuation scholars that economic value is not intrinsic to objects or practices, but rather is the outcome of a value-adding process itself mediated by various tools and technologies (Doganova and Muniesa 2015; Muniesa 2012, 2014; Ortiz 2013). Economic value is produced and performed through calculative devices such as business models (Doganova and Muniesa 2015; Doganova and Eyquem-Renault 2009) and research methods (Campbell et al. 2017; Helgesson et al. 2016) as well as the recruitment of advisors and consultants whose expertise and labor are critical to the enactment of new regimes of value (Bessy and Chauvin 2013; Birch and Tyfield 2013; Lohmann 2005; Pollard et al. 2008; Pike and Pollard 2010; Randalls 2010). It is these practices and devices, and the work of what Barman (2016) refers to as "value entrepreneurs," which underlies the process of assetization as the creation of a form of value which can be monetized, capitalized, and translated into "property that yields an income stream" (Birch 2017, 468).

Informed by this scholarship, the remainder of this chapter examines the extent to which SIBs may indeed be viewed as a type of asset and product of a distinct process of assetization. In light of the valuation challenges noted above, the focus is not simply on SIBs as further evidence that "almost anything can be turned into an asset" (Birch and Muniesa, this volume), but also on the barriers to this process and thus the limits to the assetization of social services. This analysis is informed by the results of a three-year study of SIBs and the funding of social services in Canada, the US, and the UK. The research consisted of documentary research and 195 semi-structured interviews with the various actors populating the SIB economy, including investors, government officials, service providers, and SIB intermediaries and advisors (hereafter referred to as SIB specialists). The interviews were conducted between May 2016 and July 2018 primarily in the cities of Toronto, Boston, and London, the epicenters of the Canadian, US, and UK SIB markets. These confidential interviews were then transcribed and analyzed using a coding system developed and refined during the course of the study.

Turning Social Problems into Assets: SIBs and the Valuation of Nonprofit Work

Before we can broach the question of the forms of valuation that underlie SIBs, we first need to ask a seemingly basic but surprisingly nuanced question: What is a SIB? The very term *social impact bond* is a misnomer.[6] Rather than a fixed return with no (or limited) risk, a SIB is essentially a working capital loan with a variable interest rate thus combining both debt and equity-like features. These loans have been structured in a variety of different ways. One of the more common options includes a special purpose vehicle (SPV) which holds and manages the contract. Investors thus either make an equity investment or a loan to the SPV, and the managers of the SPV (usually an intermediary or fund manager) then contract with the provider and government payor while providing governance and oversight through the life of the project. Other structures include direct loans between investors and providers, with the latter assuming a much greater share of the risk.

Beyond the selection of the contracting structure, the essential element of any SIB is the ability to translate the social impact produced by a provider or program into an economic value which can then be monetized and capitalized. As explained by one prominent figure in the field, this hinges on the creation of cash flow from services that do not in and of themselves generate any revenue;

> [The SIB model] says that... not only does social impact have economic value, it can be monetized. You can monetize that economic value and create a cash flow from it. I know from my days in investment banking, you give me a cash flow that I can identify, lock up, and secure, I can finance anything. And that's why there's so much excitement about things like SIBs. (Pinakiewicz 2014)

Given that the state is the source of this cash flow, the starting point for most projects is identifying public services that have high costs and where existing approaches have had little success. SIB designers are thus interested in "monetizable social ills... areas where there is still a very significant, often more significant kind of acute care cost" (Canadian SIB specialist #4). For example, corrections is "an inviting investment opportunity because of large budgets, the pressing demand to reduce spiraling costs, and high recidivism rates ripe for reduction" (Cullen 2013, 355). The same is true of homelessness, child welfare, and unemployment. Within these issue areas, the objective is to identify specific subpopulations that are deemed to be especially costly. These are the "high utilizers," those at high risk of reoffending (Third Sector 2013), or the chronically homeless who have frequent contact with emergency health and criminal justice systems (Segal et al. 2018).

Having defined these populations and determined the costs of existing services, SIB developers then seek to identify interventions shown to produce improvements in the outcomes for these groups. These improvements must be of sufficient size and scale to warrant government interest and to produce the requisite financial returns. The very notion of outcomes signals a fundamental shift in the way that nonprofits are evaluated. Historically, agencies have been compensated on the basis of outputs, transaction-based indicators of agency activity and contact with client groups—for example, the number of people housed in a homeless shelter. More recently, the emphasis has shifted toward payment on outcomes, understood as discernible changes or improvements in an underlying social condition. Whereas outputs are retrospective, outcomes are future-oriented and grounded in a preventative logic. In the case of homelessness, the transition to more permanent housing. For an employment program, not simply completing job training but securing full-time employment. Thus SIBs are rooted in a "shift away from a revenue model that funds *outputs*... to revenue for organizations on the basis of the *positive value of the outcomes* they achieve" (Boggild 2013). In this respect, the very notion of an outcome already reflects a form of valuation work and is central to both the construction of social value as a product of nonprofit effort and the monetization of that value in the form of prospective cost savings—that is, calculations of what these outcomes are worth.

The final element of a SIB is determining investor risks and returns and thus pricing the transaction which involves the construction of a financial model. As noted by Doganova and Muniesa (2015, 120), models are central to the practices of valuation and capitalization that underlie the "asset-becoming process." In the case of SIBs, these financial models input variables such as outcomes (as well as indicators of outcomes such as referrals and enrollments), costs, and returns, and then allow investors to test different scenarios including, for example, how variations in referral rates are likely to impact outcomes and returns, "so that you can actually say, okay, well, let's do a sensitivity around what would our returns be at different levels of performance" (Canadian SIB specialist #4). As explained by a US respondent, "We generally will develop a financial model that the lenders can use and play around with and make their own determination of the risks they are taking on and what their scenario analysis can be" (US SIB specialist #9). "Playing around" with the model allows investors to get comfortable with the SIB while subjecting social programming to the rigors of financial analysis, what Cooper et al. (2016, 73) describe in the context of the London rough-sleeping SIBs as a "layering of a grid of economic analysis (discounted cash flows, interest rates, cost allocation methods, risk assessments) onto a social field." The result of this financial modeling is the creation of a distinct "calculative space" (Callon and Muniesa 2005, 171), one which contains different hypothetical worlds and allows for different translations between social (outcomes), economic (cost savings), and financial (returns) value.

However, the significance of these models does not end with the execution of the deal. They also play a central role in what respondents suggested was the most critical aspect of any SIB: post-execution performance management. A key responsibility of SIB specialists is to assess indicators and outcomes in as close to real time as possible, making course corrections where performance falls below expectations. As explained by one investor in the context of homelessness SIBs, "in the first year if you don't get a certain number of your cohort usually into the first stage of temporary accommodation, you're never going to reach the remaining outcomes. And therefore in the first year... you really have to focus on those outcomes because if you don't get those, *nothing else matters*" (UK SIB investor #6, emphasis added). Indeed, it is widely acknowledged that programs will struggle in the early going and that many of the key assumptions will turn out to be wrong (Fraser et al. 2018b), "whatever you launch never works or some aspects of it never work as well as you hope so you always have to change some stuff" (UK SIB specialist #5). The "stuff" that needs to be changed is often dictated by the financial model itself as "investors are always constantly recalculating their returns and asking for little bits of extra work to be done" (UK provider #4). Additional resources may be required, or there may be staff changes. For SIBs, performance management is thus "where the magic happens" (UK SIB specialist #10), the point where reality is brought back in line with the model.

Following this discussion, it would appear that SIBs are indeed indicative of the transformation of social services into assets. Fundamentally, they allow for the conversion of improvements in human capital into future cash flows and thus investor returns (Cooper et al. 2016), a way to extract economic and financial value from the social value produced by nonprofits. Through this process, they share many of the defining features of assets (Birch and Muniesa, this volume). SIBs are legal contracts. They provide a monopoly over a defined service, locking-in specific programs and providers (Neyland 2018), and they allow for the extraction of rents based on exclusive rights to the value (and cost savings) produced by a larger service ecosystem with SIB providers receiving compensation despite having to rely on the services and supports of other organizations.[7] And the value of SIB investments is constantly monitored and managed through the work of performance management with the element of control often critical to the ability of SIB specialists to raise capital: "It's easier... to pull the fundraise if they can also convince the funders that they have the contracting authority and project management authority and ability to kind of change the deal if it's not looking as though it's successful" (US SIB Specialist #17). SIBs are thus perfect expressions of "both structural and sociotechnical power" (Birch and Muniesa, this volume), the former reflected in the very contracting of the SIB as an expression of state authority, and the latter in the use of data, evaluation, cost accounting, management systems, and financial models to render nonprofit work visible as a form of social, economic, and financial value.

And yet, SIBs also diverge from most other assets in several key respects. Much to the chagrin of early advocates, there is no secondary market. As a result, SIBs lack one of the essential attributes of assets: liquidity. The fact that they cannot be traded also means there is no opportunity to create value through the "relay process" (Birch 2017, 473) where "financial actors seek to add value to their financial investments before passing it onto someone else." And, with a predetermined lifespan, they cannot be held as property and accrue value over time. In this respect, the management process described above is more indicative of safeguarding rather than adding value. Thus, SIBs may be assets, but they are rather strange assets mirroring conventional investment propositions in some respects but departing quite significantly in others. This slippage and ambiguity may be one reason why SIB markets have struggled to engage more return-motivated investors. However, my conversations with SIB specialists revealed a series of additional challenges that emerge from the very practices of valuation and assetization described above and which speak to competing notions and interpretations of "value." With the design process ultimately much more fraught than commonly recognized, these valuation struggles may be the difference between turning social services into assets and creating a viable and sustainable market for those assets.

Valuation Struggles and Controversies

Despite public expressions of optimism, those working on the frontlines of the SIB market in the UK, US, and Canada expressed frustration with the slow growth of the market and a sense of uncertainty and in some cases downright skepticism as to the future of SIBs: "When I first started in social investment early in 2011, I wouldn't have quite put it as boldly as this, but I could see a social impact bond on every street corner.... But actually now I don't think it'll happen. And I think that the bubble has burst already. And I think this will fizzle out" (UK provider #2). In the words of another respondent, "The external view is it's swans on water. But there's some desperate paddling going on. And there are so few of these SIBs still in this country" (UK provider #10). One senior member of the US PFS space suggested that the PFS pipeline has "dried up significantly" and predicted that the US market would either collapse or become a "boutique" market (US SIB specialist #12). And yet, what is most interesting is the reasons cited for these struggles. Although technical barriers and deficits in government and provider capacity and expertise figured prominently in the list of headaches, respondents also pointed to the challenge of valuing these transactions and aligning the different players around a common value proposition. Three key valuation challenges emerged as being especially significant.

Outcomes versus Impact

The first of these challenges involves the question of how program effects are to be defined and evaluated. The idea in the original model was that the outcomes of SIB groups would be compared to a matched sample with the government paying on the basis of the relative improvement in the outcomes of the former relative to the latter. Thus, payment was to be contingent on a measure of *net benefit*. In the case of the Peterborough SIB, this meant that reductions in recidivism were defined relative to a comparison group of national offenders possessing similar characteristics, a quasi-experimental methodology rooted in propensity score matching. From a government perspective, this approach is valuable in controlling for deadweight—the risk of paying for outcomes that would have happened anyway.

Informed by the larger movement around evidence-based policymaking, many of the early projects in the US adopted an even more "rigorous" standard of evaluation basing program outcomes and investor payments on the results of randomized controlled trials (RCTs). For many in the evaluation community, RCTs are the gold standard of program evaluation. They are deemed uniquely capable of not only controlling for deadweight but also addressing the problem of attribution and causality—that is, the extent to which programs are singularly responsible for observed changes in outcomes. Here it is the element of randomization that is key as the random assignment of individuals to intervention and control groups is believed to allow for the control of extraneous and confounding variables thus distinguishing between genuinely causal and merely correlational effects. It is this epistemic virtue of causality that is central to the exalted status of the RCT as an arbiter of effective and proven programs. With many of those working in the US PFS space well-schooled in the merits of RCTs, including the Arnold Foundation and advisors such as the Urban Institute and Harvard's Government Performance Lab, this method was quickly adopted as the standard for PFS deals and the basis for determining program outcomes and investor returns (Milner and Walsh 2016). Thus, the economic value of the SIB is tied to the epistemic virtues of the RCT as a particular type of "counterfactual display" (Ehrenstein and Muniesa 2013, 162).

Not surprisingly, this use of RCTs (and even quasi-experiments) in the SIB context has received extensive criticism and is the subject of an increasingly heated debate (Savell and Heady 2016). Among the concerns cited by intermediaries and investors is that RCTs are costly and can be logistically challenging adding to the complexity and timelines for SIB deals. This includes the need to generate sufficient referrals to populate program and control groups, and the requirement that programs are large enough (100 to 200 is often used as a benchmark) to yield statistically valid results (Bolton and Savell 2010; Fox and Albertson 2011; MaRS 2013). From an investor perspective, RCTs also introduce a new form of risk—evaluation risk. "[The RCT] adds a ... risk that in the early days we didn't think of, which is evaluation risk, which basically is the risk the structure and process of the evaluation will actually impact the results that are observed" (US investor #1). This includes the fact that investor returns are subject to the vagaries of statistical technique and are dependent on producing not only positive but statistically significant results. "So you have government saying ... you can't pay unless you have a statistically significant result, so you need 95 percent" (US SIB specialist #7). This evaluation risk is especially challenging given that it is "divorced from any actuarial basis" (US government official #1) thus introducing uncertainty rather than simply risk into the transaction. As noted by a UK-based investor, "paying against outcomes linked to an RCT or an experimental or quasi-experimental evaluation ... takes what is already quite a risky proposition with a lot of factors that you can't really control and it squares them" (UK SIB specialist #10).

For intermediaries and fund managers, RCTs (and quasi-experiments) come with another significant downside. In the absence of regular data on the control or comparison group, it is difficult to manage toward specific payment outcomes thus impeding the forms of performance management that are so central to SIBs. "It's quite difficult to deliver when you don't know how well you're doing. We didn't know how well we were doing until retrospectively we were measured" (UK SIB specialist #4). In reiterating the importance of performance management, one respondent likened RCTs to a "black box." "It's really about performance management and so getting that real-time feedback is really important. And the RCT doesn't lend itself to that. It's like a black box that you open up one day" (US investor

#1). This performance management challenge was actually cited by several respondents as a key factor in the movement of the UK market away from quasi-experiments and live counterfactuals (Carter et al. 2018; Fraser et al. 2018b). In fact, out of the forty-plus SIBs launched in the UK between 2010 and 2018, only the first two (Peterborough and Essex) draw from a live comparison group. The vast majority are instead based on a rate card in which government develops a menu of pre-priced outcomes and providers are compensated for each individual outcome they achieve. More recently, the US has also moved toward this rate card model and away from RCTs, much to the chagrin of the following respondent who saw this as a response to weaker than expected results from the early slate of RCT-based deals,

> Some of the earlier deals were coming online and there was sort of hushed awareness that the results from some of these deals were not as positive as had been hoped. So, instead of having a collective discussion about why and what that meant for iteration, some of the senior folks from all of the intermediaries determined that the best way to go was…to water down the evaluation design. That there was too much risk involved in these deals. So they started going away collectively from RCTs. And I think there's issues with RCTs in certain settings. But going away from RCTs and even quasi-experimental designs, going actually back to some pre/post measures and discussions of parachute clauses you could put into contracts such that you could break contracts fairly easily if it's not getting results or inflows of participants. That was very concerning to me. (US SIB specialist #17)

The issue here is not whether RCTs are good or bad. There is an extensive literature that has rightly questioned the merits of RCTs as an aspect of program evaluation (Donovan 2018) as well as the ability to attribute changes in complex social conditions to a single program or provider (Lowe 2013; Lowe and Wilson 2017). The point is that these competing arguments around the merits of RCTs as a feature of SIB design are rooted in different interpretations of the "value" of these investments and the rules of counterfactual display. From the perspective of government, "value" depends on net benefit and the confidence that they are paying investors for true program effects validated using the most rigorous measures possible. For investors and SIB specialists, RCTs are a source of uncertainty and a barrier to performance management. In advocating for the move away from RCTs, practitioners have invoked alternative notions of value suggesting that payment based on outcomes is itself a significant improvement over existing government practice which remains focused on activity-based indictors. Some outcomes, such as moving the homeless into accommodation, are also deemed to be inherently good rendering comparisons unnecessary: "Because *it is a good*... if you're chronically homeless and we have you in housing for six months, we don't need to compare that to anything because we know that by definition treatment as usual was not solving that problem, that's why you're chronically homeless" (US SIB specialist #7). These contrasting views continue to represent a key point of tension in the field with investors and government often working at cross-purposes.

Cashable Savings versus Value for Money

A second valuation challenge underlying SIBs concerns the ability of the government to generate and realize the cost savings from these transactions. Another promise of the original SIB model was that investments in preventative programs would yield not only savings to government, but savings that were cashable (i.e., manifested as savings in specific budget lines) and could thus be used to pay investor returns. The notion that SIBs yield cash flow for government in this way is central to the monetization of program outcomes as well as the government value proposition as this allows not only investors but government to effectively cash out of these deals. However, in developing the first wave of projects, SIB specialists quickly realized that there were significant barriers to this view of the liquidity of public capital (Disley et al. 2011; Fraser et al. 2018b; McKay 2013).

Beyond the more general challenge of how to allocate cost savings to individual government payors, particularly in cases where savings may accrue to multiple agencies as well as levels of government, one key barrier to the cashability of cost savings is the nature of public sector costs, which tend to be fixed rather than variable. In order to produce meaningful cost savings in the context of a SIB, the improvement in outcomes would need to be of sufficient size, scale, and duration to allow for reductions in these fixed costs. This is easier said than done. A perfect example of this challenge is reoffending programs which were identified early on as promising candidates for SIBs given the high costs of police, courts, and corrections. The difficulty is that most of these costs involve buildings and staff. Even a significant reduction (e.g., 20 percent) in "bed days"—the measure typically used in these transactions—would not be enough to close the wing of a prison, the point where real savings would start to accumulate (McKay 2013). Closing prisons and laying off staff may also create additional political liabilities. "You've then got to lay off large numbers of statutory workers which is very expensive and the unions get very involved and it becomes difficult" (UK provider #9). As with public services more generally, there is also the challenge that any additional capacity will simply be backfilled with latent demand. "How often are the cashable savings realized? We try not to talk about them. Because you can guarantee that within public services there is latent demand" (UK SIB specialist #10).

Given these difficulties, many in the SIB space have moved away from a strict cashable savings approach: "If you dig deep, cost savings is the rhetoric. Even if you reduce recidivism, you don't really save much money. It's just a way to talk about it. At the end of the day it's about getting value for the money that is being spent" (US SIB specialist #15). One way of getting greater value for money is to focus not on new spending streams (based on the rationale that they will enable the prevention of future costs), but rather on reducing costs and achieving greater efficiencies in existing spending. "We think there is also a substantial opportunity in improving existing services, i.e., in helping commissioners achieve better value in situations where they already have a targeted spend (either by achieving better outcomes for the same spend, the same outcomes for less spend, or more outcomes for more spend but at a lower cost per outcome)" (Bridges Ventures 2016, 8). A US-based respondent provided the following rationale for this shift in approach: "It's an efficiency claim for savings rather than a cashable saving.... So if you can enable the current, the existing spending streams to be PFS enabled and therefore get more efficiency out of them, then you don't have to make that same savings argument that we are saving money in the long-run" (US SIB specialist #7). There have also been attempts to incorporate broader notions of "public value" (Kohli et al. 2015). In the case of criminal justice, savings could accrue not only from marginal cost savings but also reduced costs to victims, including medical costs, lost earnings, and psychological pain and suffering (Fogel et al. 2017). Indeed, victim impact was included in the projected savings for the Peterborough (Disley et al. 2011) and the New York State (GAO 2015) SIBs.

From the perspective of government, these types of savings may be even more challenging to calculate and may rely on potentially tenuous connections between near-term program outcomes and longer-term impacts (Heinrich and Kabourek 2018, 8–9). As Fraser et al. (2018a, 16) conclude from their review of the available SIB literature, "Many of the savings in SIB schemes appear to be based on hypothetical rather than real cost reductions, are complicated to calculate and in the absence of (quasi) experimental impact evaluations, hard to attribute." This is indicative of what Neyland (2018) describes as a form of "calculative asymmetry" between government and investors with the former less able to project and model future outcomes and savings. The larger point here is that there is a disconnect not only in calculative competency and capacity but also the type of value that is created and imagined in these deals. For investors, value is well defined with specific outcomes equated with set returns. For government, savings and notions of value remain much more hypothetical and promissory (Martin 2015) with the outcomes on which they are required to pay often lacking any direct connection to the savings and forms of value associated with these deals thus further eroding the government value proposition.

Risk versus Return

While benefiting from greater sophistication in financial modeling and certainty in terms of the payoff if program outcomes are realized, SIBs are still challenging for investors. As noted earlier, they are unconventional assets. Absent a secondary market, they have virtually no liquidity, and with returns capped, there is little speculative upside. SIBs also have distinct attributes that make them especially risky. The mere fact that they are not rooted in a traditional type of physical asset, principally real estate, is a source of concern particularly for those in the community development space such as Community Development Financial Institutions (CDFIs) and Community Reinvestment Act (CRA) banks.[8] These organizations, which are used to dealing with both financial and social returns, were initially viewed as perfect candidates for SIBs and a bridge to more commercial investors (Godeke and Resner 2012). However, they have tended to view SIBs as overly risky given their departures from these traditional asset classes. "When you dig into the community development industry in the USA, be it through the CDFIs or the CRA banks, most of it is funding real assets and hard assets which they can get their heads around. And this was like kind of a challenge for them to figure out. What's the risk/return profile of this structure?" (US advisor #3).

SIBs are also challenging given that they are rooted in the world of social programming and are dependent on projected changes in human behavior. Even for the most sophisticated investors, they can thus be difficult to diligence.

> What does it mean to diligence a social service outcome? That's very different than what their internal investment committee are used to reviewing. It's very different if they have an investment manager, what they are used to doing. It can be very different if they are a regulated entity to make sense to their internal compliance folks. What does it mean that your payment is contingent on human behavior? (US SIB specialist #9)

Invariably it means that investors will have to engage with a social science evidence base and confront a series of new and distinct forms of risk (GAO 2015; Godeke and Resner 2012; Social Finance 2012). In addition to the aforementioned evaluation risk, there is policy risk, "the risk that a government initiates a policy change that prevents a PFS project from operating as initially intended. A policy change could disrupt a service provider's program delivery, putting the achievement of outcomes—and investors' investment—at risk" (GAO 2015, 35). This begs the question of who should "own" this risk, a point the following respondent illustrated in reference to a criminal justice program:

> Who should own the fact that either government could change the rules and not send people to jail as much, or that something could change in the real world, the opioid epidemic, or an increase in gun violence, or an economic drop or an economic improvement? I'm not sure who should own the risk for that when you're doing a five, six, seven year study. (US SIB specialist #7)

There is also appropriation risk, the risk of entering into a long-term contract with government and the possibility that these commitments might not be honored by future administrations.

From an investor perspective, SIBs thus possess a number of unattractive properties. They have the risk profile of equity investments but the structure and returns of debt. They require a form of expertise and type of valuation work that is foreign to most investors, with the small size of individual deals and limited deal flow providing few incentives to develop these capabilities. They are rooted in new forms of risk that are challenging to define and price. And these transactions are often bespoke and resistant to standardization with the distinctly local nature of social problems requiring that the valuation process be undertaken anew with each deal. While the UK market has been sheltered to a large extent by a steady supply of investment capital through Big Society Capital, these features help to explain the struggles reported by US respondents in engaging investors. "You have funders saying look at the high risk. . . . There's not one project in the USA or around the world that is not extremely high risk" (US SIB specialist #4). Reflecting this risk, one respondent reported that a major US intermediary actually "had to shut down a handful of very large deals because they couldn't do the fundraise for the deal" (US SIB specialist #17), with another respondent commenting that, "I don't think there's as much money out there that is as interested in taking on the risk as people thought there was" (US SIB specialist #7). Moreover, efforts to make these investments more attractive to investors—for example, by providing earlier repayments based on outputs (e.g., enrollment) rather than outcomes—further erode the government value proposition. One respondent described how a proposal for early payment based on enrollments rather than outcomes was a key sticking point in negotiations with government who "[held] hard and fast to 'This is called PFS. Success is showing somebody that they did something and so I'm not going to pay you for an enrollment payment'" (US provider #8)—although they did capitulate under the pressure to get the deal done. This further illustrates not only the valuation challenges that underlie SIBs but also the fundamental misalignment of government and investor interests.

Conclusion

Viewed from a distance, SIBs would appear to confirm fears about the spread of finance into more and more aspects of social life. The fact that marginalized populations—offenders released from prison, the chronically homeless, the unemployed—are being transformed into investment propositions is thoroughly in line with the dystopian vision of finance. And yet, while this view may be faithful to the logic of SIBs and the aspirations of proponents, a closer look reveals a market that has struggled to take hold and to translate this vision into reality. The return-motivated investors prized by advocates and feared by critics have largely failed to materialize, and the main drivers of the market are not financiers but a small group of advisors and consultants backed by government and philanthropy. Far from a "robust growth sector" (Lake 2016, 14), the market has experienced slower growth than expected and may be more accurately viewed as a small, niche market (Williams 2019).

Informed by the results of an extended analysis of SIBs in the UK, US, and Canada, this chapter has argued that these realities of the SIB market are difficult to square with the financialization narrative featured in many critical accounts and that thinking about SIBs in terms of an asset (and process of assetization) helps to capture both their attributes and limitations as a type of investment proposition. The analytic of assetization focuses attention on the essential question of how SIBs are valued and the specific practices, sociotechnical knowledges, and forms of valuation work through which social programs are reconfigured as outcomes and translated into claims to economic (savings) and financial (returns) value. This involves the monetization not necessarily of programs themselves, but of the knowledge of these programs constituted through devices such as RCTs, cost-benefit analyses, and financial modeling. In addition to this reconfiguration of knowledges, practices, and relations, SIBs share several other features of assets as outlined in this volume. They are the product of state fiat, as are all legal contracts, and they represent a form of monopoly in which exclusive rights are granted to the value and rents produced by a larger ecosystem of providers. Their value is also subject to ongoing management even after deals have been signed, part of a "contractual [regime] of governance" (Birch and Muniesa, this volume).

At the same time, SIBs are not like most other assets. There is no secondary market, their revenues are time limited, and their upside is capped by the terms of the contract (see Nadai and Cointe, this volume). More importantly, the valuation practices at the heart of SIBs are invariably somewhat fraught. Reflected in each of the distinct challenges examined in this chapter, including the nature and terms of counterfactual display, the ability of the government to realize the savings and capture the value from these transactions, and the disconnect between risk and return and investor uncertainty around how to assess and price these deals, the effort to value SIBs has been undercut by competing interests and conflicting interpretations of how to value nonprofit work. There are also inherent tensions between the complex and messy realities of social services and the valuation frames and repertories used by both government and investors. The notion of outcomes-based savings is not easily aligned with government budgets or the rules of public finance, while investors struggle with the uncertainty and unpredictability that comes with investing in putative changes in human behavior. Ultimately, it is these valuation challenges and dilemmas which help to account for the struggles to execute these deals and to grow the market. SIB specialists may have been successful in turning social services into an asset, but it is a strange asset with features that are less than appealing to government and investors alike. Viewing SIBs through the lens of assetization thus provides a very different perspective, one that is more consistent with the realities of the market but which has been largely overlooked in the literature to date.

There are two further implications that follow from this analysis. First, the case of SIBs reinforces the importance of focusing not only on the process of turning things into assets but also the challenges and limits encountered as part of this work. As noted by Birch and Muniesa (this volume), "Examining how things are turned into assets means understanding how assets are then maintained or challenged as such." These challenges include the micropolitics that often underlie forms of assetization and the fact that devices such as business and financial models are not only sources of alignment and coordination (Doganova and Muniesa 2015; Doganova and Eyquem-Renault 2009; MacKenzie and Spears 2014), but also points of conflict and division. The case of SIBs also suggests that there is a subtle distinction to be made between turning things into assets and developing a viable and sustainable market for those assets, the latter hinging on the ability to engage key parties over the longer term and as part of a shared future vision for the market.

Second, SIBs provide an opportunity to examine processes of assetization as they relate to the world of public versus simply private finance. As noted by Birch and Muniesa (this volume), while the analysis of assetization usually involves a focus on the appropriation of value by private investors and thus the "expansion of private ownership claims over more and more aspects of our lives," what may be most noteworthy about SIBs is their role in reconfiguring nonprofit work as a type of *public* asset (see Milyaeva and Neyland, this volume). This follows from the recent evolution of the SIB market, particularly in the US, where several advisory firms have shifted their focus away from the private capital aspect of SIBs and are seeking to work directly with governments to improve the way they contract with the nonprofit sector. Utilizing the tools and lessons gleaned from their SIB work, the emphasis is on reengineering existing spending streams using data analysis to identify inefficiencies in services, and performance management to exhort providers to address these inefficiencies and improve outcomes. This approach gets around some of the valuation challenges associated with having to engage investors, while still requiring nonprofits to "do the hard work of quantifying their value—defining the outcomes that they influence and estimating the fiscal and social value of those outcomes to key funders" (Segal et al. 2016, 36). In mandating this valuation work and building contracts around the resulting outcomes, it is government that is ultimately taking on the role of investor extracting a type of public rent from the nonprofit sector. By focusing almost exclusively on the financial aspects of SIBs, and the dystopian vision of financialization, commentators have largely overlooked these developments around outcomes-based funding which are likely to have a much more significant and enduring impact on the nonprofit sector. All of this offers a slightly different perspective on the "assetization of public policy" (Birch and Muniesa, this volume) as well as on the nature of assets themselves.

Notes

1. In the US, this model is referred to as "Pay-for-Success" (PFS) while in Australia "Social Benefit Bonds" is the preferred term. For the sake of clarity, "social impact bond" will be used throughout this chapter except where specific reference is being made to the US context.

2. The SIB model is also making inroads in the Global South, primarily in a development context, through its close cousin the Development Impact Bond (DIB).

3. For example, Bloomberg Philanthropies, the foundation established by former New York City mayor Michael Bloomberg, provided a guarantee of $7.2 million for Goldman Sach's $9.6 million investment in the Rikers Island SIB, the first deal in the US.

4. Several US SIB deals have drawn from capital stacks in which foundations have served as subordinate investors allowing more return-motivated investors to come in as senior capital receiving higher returns, earlier payouts, and less risk.

5. In fact, two firms (one based in Canada and the other in the UK) were unable to survive and were absorbed by other players in the market.

6. This misleading terminology created immediate challenges in marketing the product particularly in the North American context with US practitioners quickly adopting the alternative moniker of "Pay-for-Success."

7. The extraction of value through relationships with other service providers is an explicit feature of many of SIB programs which are rooted in "navigator" or "link worker" models where the whole purpose of the intervention is to link clients to other services.

8. Both CDFIs and CRA banks provide credit and financial services to underserved communities in the USA and are often involved in financing economic and community development initiatives.

References

Arena, M., Bengo, I., Calderini, M., and Chiodo, V. 2016. Social Impact Bonds: Blockbuster or Flash in a Pan? *International Journal of Public Administration* 39 (12): 927–939.

Barman, E. 2016. *Caring Capitalism: The Meaning and Measure of Social Value*. Cambridge: Cambridge University Press.

Bessy, C., and Chauvin, P. 2013. The Power of Market Intermediaries: From Information to Valuation Processes. *Valuation Studies* 1 (1): 83–117.

Birch, K. 2017. Rethinking Value in the Bio-Economy: Finance, Assetization, and the Management of Value. *Science, Technology, and Human Values* 42 (3): 460–490.

Birch, K., and Tyfield, D. 2013. Theorizing the Bioeconomy: Biovalue, Biocapital, Bioeconomics or... What? *Science, Technology, and Human Values* 38 (3): 299–327.

Boggild, L. 2013. Outcome Finance Webinar for the J. W. McConnell Foundation's Innoweave Program. December.

Bolton, E., and Savell, L. 2010. *Towards a New Social Economy: Blended Value Creation Through Social Impact Bonds*. London: Social Finance.

Bridges Ventures 2016. *Better Outcomes, Better Value: The Evolution of Social Impact Bonds in the UK*. March, 2016.

Callon, M., and Muniesa, F. 2005. Peripheral Vision: Economic Markets as Calculative Collective Devices. *Organization Studies* 26 (8): 1229–1250.

Campbell, P., Cox, T., and O'Brien, D. 2017. The Social Life of Measurement: How Methods Have Shaped the Idea of Culture in Urban Regeneration. *Journal of Cultural Economy* 10 (1): 49–62.

Carter, E., FitzGerald, C., Dixon, R., Economy, C., Hameed, T., and Airoldi, M. 2018. *Building the Tools for Public Services to Secure Better Outcomes: Collaboration, Prevention, Innovation*, July 18, 2018. Oxford: Government Outcomes Lab.

Cooper, C., Graham, C., and Himick, D. 2016. Social Impact Bonds: The Securitization of the Homeless. *Accounting, Organizations, and Society* 55: 63–82.

Cullen, F. 2013. Rehabilitation: Beyond Nothing Works. *Crime and Justice* 42: 299–376.

Dear, A., et al. 2016. *Social Impact Bonds: The Early Years*. July. Boston: Social Finance.

Dey, C., and Gibbon, J. 2018. New Development: Private Finance Over Public Good? Questioning the Value of Impact Bonds. *Public Money and Management* July: 375–378.

Disley, E., Rubin, J., Scraggs, E., Burrowes, N., and Culley, D.M. 2011. *Lessons Learned from the Planning and Early Implementation of the Social Impact Bond at HMP Peterborough.* Cambridge: RAND Europe.

Doganova, L., and Eyquem-Renault, M. 2009. What Do Business Models Do? Innovation Devices in Technology Entrepreneurship. *Research Policy* 38 (10): 1559–70.

Doganova, L., and Muniesa, F. 2015. Capitalization Devices: Business Models and the Renewal of Markets. In *Making Things Valuable*, edited by M. Kornberger, L. Justesen, A. Madsen, and J. Mouritsen, 109–125. Oxford: Oxford University Press.

Donovan, K. P. 2018. The Rise of the Randomistas: On the Experimental Turn in International Aid. *Economy and Society* 47 (1): 27–58.

Dowling, E. 2017. In the Wake of Austerity Social Impact Bonds and the Financialisation of the Welfare State in Britain. *New Political Economy* 22 (3): 294–310.

Dowling, E., and Harvie, D. 2014. Harnessing the Social: State, Crisis, and the (Big) Society. *Sociology* 48 (5): 869–886.

Edmiston, D., and Nicholls, A. 2018. Social Impact Bonds: The Role of Private Capital in Outcomes-Based Commissioning. *Journal of Social Policy* 47: 57–76.

Ehrenstein, V., and Muniesa, F. 2013. The Conditional Sink: Counterfactual Display in the Valuation of a Carbon Offsetting Reforestation Project. *Valuation Studies* 1 (2): 161–188.

Floyd, D. 2017. *Social Impact Bonds: An Overview of the Global Market for Commissioners and Policymakers.* April. London: Centre for Public Impact.

Fogel, A., Shumway, J., and Udpa, A. 2017. *Challenging the Status Quo: Philadelphia's Exploration of Pay for Success.* Boston: Social Finance.

Fox, C., and Albertson, K. 2011. Payment by Results and Social Impact Bonds in the Criminal Justice Sector: New Challenges for the Concept of Evidence-Based Policy? *Criminology and Criminal Justice* 11 (5): 395–413.

Fraser, A., Tan, S., Lagarde, M., and Mays, N. 2018a. Narratives of Promise, Narratives of Caution: A Review of the Literature on Social Impact Bonds. *Social Policy and Administration* 52 (1): 4–28.

Fraser, A., Tan, S., Kruithof, K., Sim, M., Disley, E., Giacomantonio, C., Lagarde, M., and Mays, N. 2018b. *Evaluation of the Social Impact Bond Trailblazers in Health and Social Care. Final Report.* PIRU Publication 2018–23.

Government Accountability Office. 2015. *PFS: Collaboration Among Federal Agencies Would Be Helpful as Governments Explore New Financing Mechanisms*. Report to the Chairman, Committee on the Budget, US Senate. GAO-15–646.

Giacomantonio, C. 2017. Grant-Maximizing, but Not Money-Making: A Simple Decision-Tree Analysis for Social Impact Bonds. *Journal of Social Entrepreneurship* 8 (1): 47–66.

Godeke, S., and Resner, L. 2012. *Building a Healthy and Sustainable Social Impact Bond Market: The Investor Landscape*. New York: Godeke Consulting.

Heinrich, C., and Kabourek, S. 2018. *Social Impact Bonds (or Something Like That): Are Viable and Sustainable Models Emerging?* Working Paper. September 2, 2018.

Helgesson, C. F., Lee, F., and Linden, L. 2016. Valuations of Experimental Designs in Proteomic Biomarker Experiments and Traditional Randomized Controlled Trials. *Journal of Cultural Economy* 9 (2): 157–172.

Joy, M., and Shields, J. 2013. Social Impact Bonds: The Next Phase of Third Sector Marketization? *Canadian Journal of Nonprofit and Social Economy Research* 4 (2): 39–55.

Kohli, J., Golden, M. Coletti, J., and Bo'sher, L. 2015. *From Cashable Savings to Public Value: Pricing Program Outcomes in Pay for Success*. Washington, DC: Center for American Progress.

Lake, R. 2016. The Subordination of Urban Policy in the Time of Financialization. In *Urban Social Policy in the Time of Obama*, edited by J. DeFillippis, 45–64. Minneapolis: University of Minnesota Press.

Lohmann, L. 2005. Marketing and Making Carbon Dumps: Commodification, Calculation and Counterfactuals in Climate Change Mitigation. *Science as Culture* 14 (3): 203–235.

Lowe, T. 2013. New Development: The Paradox of Outcomes—The More We Measure, the Less We Understand. *Public Money and Management* 33 (3): 213–216.

Lowe, T., and Wilson, R. 2017. Playing the Game of Outcomes-Based Performance Management: Is Gamesmanship Inevitable? Evidence from Theory and Practice. *Social Policy and Administration* 51 (7): 981–1001.

MacKenzie, D., and Spears, T. 2014. A Device for Being Able to Book P&L: The Organizational Embedding of the Gaussian Copula. *Social Studies of Science* 44 (3): 418–40.

Maier, F., and Meyer, M. 2017. Social Impact Bonds and the Perils of Aligned Interests. *Administrative Sciences* 7 (24): 1–10.

MaRS. 2013. *Social Impact Bond Technical Guide for Service Providers*. Toronto: MaRS Centre for Impact Investing.

Martin, P. 2015. Commercialising Neurofutures: Promissory Economies, Value Creation, and the Making of a New Industry. *BioSocieties* 10 (4): 422–443.

McHugh, N., Sinclair, S., Roy, M., Huckfield, L., and Donaldson, C. 2013. Social Impact Bonds: A Wolf in Sheep's Clothing? *Journal of Poverty and Social Justice* 21 (3): 247–57.

McKay, K. 2013. *Evaluating Social Impact Bonds as a New Reentry Financing Mechanism: A Case Study on Reentry Programming in Maryland.* Annapolis, MD: Department of Legislative Service, Office of Policy Analysis.

Milner, J., and Walsh, K. 2016. *Measuring Success in Pay for Success: Randomized Controlled Trials as the Starting Point.* Washington, DC: Urban Institute.

Mirowski, P. 2013. *Never Let a Serious Crisis Go to Waste: How Neoliberalism Survived the Financial Meltdown.* London: Verso Books.

Muniesa, F. 2012. A Flank Movement in the Understanding of Valuation. *Sociological Review* 59 (s1): 24–38.

Muniesa, F. 2014. *The Provoked Economy: Economic Reality and the Performative Turn.* London: Routledge.

Neyland, D. 2018. On the Transformation of Children at-Risk into an Investment Proposition: A Study of Social Impact Bonds as an Anti-Market Device. *Sociological Review* 66 (3): 492–510.

Ortiz, H. 2013. Financial Value: Economic, Moral, Political, Global. *HAU: Journal of Ethnographic Theory* 3: 64–79.

Pike, A., and Pollard, J. 2010. Economic Geographies of Financialization. *Economic Geography* 86 (1): 29–51.

Pinakiewicz, B. 2014. *Foundations for Social Impact Bonds.* March 6. Washington, DC: Aspen Institute Program on Philanthropy and Social Innovation, and Social Finance.

Pollard, J., Oldfield, J., Randalls, S., and Thomas, J. E. 2008. Firm Finances, Weather Derivatives, and Geography. *Geoforum* 39: 616–624.

Randalls, S. 2010. Weather Profits: Weather Derivatives and the Commercialization of Meteorology. *Social Studies of Science* 40 (5): 705–730.

Savell, L., and Heady, L. 2016. *Balancing Evidence and Risk: Evaluating Impact Bonds.* Social Finance and Children's Investment Fund Foundation.

Segal, J., Feng, J., Greenberg, L., and Shumway, J. 2018. *Using Pay for Success to Improve Outcomes for the Persistently Homeless in Sacramento.* Boston: Social Finance.

Segal, J., Shumway, J., and Bianchi, J. 2016. *New Tools to Amplify Impact: A Pay for Success Guide to Building Nonprofit Capacity.* Boston: Social Finance.

Sinclair, S., McHugh, N. Huckfield, L., Roy, M., and Donaldson, C. 2014. Social Impact Bonds: Shifting the Boundaries of Citizenship. In *Social Policy Review 26: Analysis and Debate in Social Policy*, edited by K. Farnsworth, Z. Irving, and M. Fender, 119–136. Bristol: Policy Press.

Social Finance. 2012. *A New Tool for Scaling Impact: How Social Impact Bonds Can Mobilize Private Capital to Advance Social Good*. Boston: Social Finance.

Third Sector. 2013. *Case Study: Preparing for a Pay for Success Opportunity*. Boston: Third Sector Capital Partners.

Warner, M. 2013. Private Finance for Public Goods: Social Impact Bonds. *Journal of Economic Policy Reform* 16 (4): 303–319.

Williams, J. 2019. *From Visions of Promise to Signs of Struggle: Exploring Social Impact Bonds and the Funding of Social Services in Canada, the US, and the UK*. Toronto: York University.

12 Conclusion: The Future of Assetization Studies

Kean Birch and Fabian Muniesa

Introduction

So, where do we end up with a book like this? As the contributions and their analytic puzzles unfold, they provide fertile terrain on which to develop the future of assetization studies in a number of ways. In particular, we think all the chapters demonstrate two things very clearly: first, that there is significant room for further studies of assets and assetization, especially as an analytical tool in our critical understanding of contemporary and future technoscientific society; and second, the more we seek to examine and unpack the particularities and peculiarities of assets and assetization, the more that wider social and political concerns start to populate this research endeavor. As such, assetization studies certainly connect with the preoccupations that characterize the multifaceted streams that fall under the label of the "critique of capitalism." But how? And what does it bring that is new to these critiques exactly?

The Future of Assetization Studies

Obviously, we cannot predict where intellectual and public trends and interest will alight next, but we do think that the "asset form" and its corollary—the "asset condition" (Muniesa et al. 2017; Birch 2018)—are an increasingly important and increasingly visible *problematic* in society. From the concerns with rising house prices and their socially destabilizing effects (see, for example, Birch 2015; Cooper 2017; Adkins et al. 2019) through the rise of an "investor gaze" and sentiment (Birch 2017a; Muniesa et al. 2017; Feher 2018) to the issue of what assets and resources are going to underpin our technoscientific futures (Haskel and Westlake 2018; Tyfield

2018; Sadowski 2019; West 2019), we are seeing a range of research avenues opening up to the question of how such diverse things are turned into the specific asset form.

Contributors to this volume outline some of this diversity and variety, certainly, but many others are doing the same work elsewhere. In their chapters, Kang, Beauvisage and Mellet, and Roy all show how supposedly ephemeral and immaterial things like knowledge, health, and data can be assetized; they thereby contribute to widening debates—and, increasingly, political concerns—about the transformation of our lives and the digital traces those lives leave behind into the "oil" of the twenty-first century—to repeat a somewhat tortuous metaphor (OECD 2012). Much of this scholarship is focusing on the issue of digital data and changing ownership rights constituting those data as assets and reconstituting ownership itself (Perzanowski and Schultz 2016; Geiger and Gross 2019; Zuboff 2019); here, some stress the need to understand the legal shifts around digital rights (e.g., Pistor 2019), while others are more focused on the conversion of personal data itself into "capital" (e.g., Sadowski 2019). As we head down the road to data-driven economies, we will no doubt see—and need—more assetization studies along these lines, picking apart the specificities of our digital lives as they are turned into assets (e.g., Nieborg and Poell 2018; West 2019).

And yet, assets are also very much bounded by their biophysical and geographical materialities, as contributors like Buier, Nadaï and Cointe, Gilbert, Braun, and Levidow show in their very distinct takes on assetization. A number of important new studies have come out recently focused specifically on the transformation of land—agricultural and urban—into a financial asset (e.g., Ducastel and Anseeuw 2017; Larder et al. 2018; Ward and Swyngedouw 2018; Ouma 2019). Notably, over a decade ago now, Leyshon and Thrift (2007) had called precisely for this sort of critical examination of "new asset geographies," but such an intellectual project has taken some time to get off the ground. That being said, there is now a growing interest in how material things are transformed into assets, especially through new forms of finance and investment logics (e.g., Bridge et al. 2019; O'Brien et al. 2019). Here there is an indication of the need to think about the particular financial and other knowledges, practices, and processes that enable specific things to be turned into assets in place, at the same as thinking about the wider capitalist (or other) logics that configure these transformations as necessary or inevitable.

Finally, this book's contributors also illustrate the fact that assetization is a boundary-crossing process: it can be both material and immaterial (Gilbert, this volume); it can entail switching between both the commodity form and the asset form—and back again (Braun, this volume); and it can be both social and antisocial, all at the same time. As Milyaeva and Neyland and Williams show, for example, assets and assetization are ways of making and remaking the social world, where attempts to address societal problems can end up legitimating an antisocial solution (also Neyland 2018). For example, capitalizing particular social costs—e.g., higher education or social care—may make sense as a way to ensure that social costs do not always fall on current generations (Muniesa et al. 2017), but in doing so they lock future generations into a techno-economic solution they have not agreed to. Examining these social futures is another research avenue that would open up the direction of assetization studies considerably, bringing in more sociological and anthropological dimensions (e.g., identity, subjectivity, sociality) to our understanding of the asset condition.

Political Challenges and Consequences

All of which brings us to the emergence and consolidation of new forms of power—*assetization power*—as a central challenge to future assetization studies. At a time when capitalism is insistently assessed in relation to inequalities in the distribution of income and resources, the analysis of the role that global financial elites play in both the organization and the appropriation of this distribution comes to the forefront (Godechot 2017). Claiming within this context, with Nitzan and Bichler (2009), that "capital *is* power" means acknowledging that turning things into assets amounts to a political process. A qualitative, constructivist examination of such a process does not take assetization for granted nor considers it as a merely technical, essentially neutral technique. The political processes it both requires *and* fosters are essential to the discussion, reflecting the underlying and necessary contingency in the transformation of things into assets (Chiapello 2015). And here we can find room to intervene politically, not only to offer new narratives but also to come up with new ways of representing the *future*, both figuratively and politically. For we would argue that at its base that is what assetization entails; namely, it is about who "owns" the future and, more importantly, *how* they end up owning it and what that means for everyone else.

This political take relates quite clearly to current preoccupations with the transformations of political sovereignty. Both the spread of neoliberal government and of economic financialization have entailed abundant reflection on the threat they pose to the democratic state (Davies 2014; Birch 2017b). The fact that attracting investors (i.e., generating a climate of confidence for the financial industry) stands as a crucial political leitmotiv all over the world indeed means that the asset form and condition are playing a crucial part in the transformations of state sovereignty (Blyth 2013; Streeck 2014; Alliez and Lazzarato 2016; Tooze 2018; Konings 2018; Feher 2018). Contemporary issues of monetary policy, financial regulation, crisis response, budgetary restraint, and fiscal strategy can be interpreted, at least in part, in light of the prominence that an investor's viewpoint has in the political identification of social problems and determination of social action. Reactions to both neoliberalism and financialization that range from protectionist measures and economic nationalism to populist proclamations may accordingly be made sense of, we believe, as tied to the problematics of assetization.

One should also remain aware of the fact that the current asset condition is not simply a technoscientific one, or an economic one, or a political one. The asset form, considered as a cultural formation too, has penetrated habits and identities in ways that can be detected in everyday life (Cooper 2017). How should we tackle the micropolitics involved in considering life in terms of investments and people in terms of assets? Our bodies, our homes, our kids, our relations are assetized, in a broad cultural sense, when we start considering their present value in the light of the future benefits they may confer. Assetization can therefore indeed be captured as a form of *subjectivation*, one that sees, for example, in the spread of a culture of self-appreciation the traces that the asset condition—or perhaps the "investee" condition—leaves (Feher 2009, 2018). Joining, at least partly, a Foucauldian tradition that sees in mundane metrics the key to particular forms of power, assetization studies can raise political challenges in the personal expressions of the asset condition.

Conclusion

We finish by emphasizing that the technoscientific micropolitics of the asset do extend, though, from the perimeter of mundane existence to the frontiers of capitalism. As the asset form features prominently as a solution to all sorts of policy problems—economic, social, environmental—vigilance is required not only so as to better assess how fit these solutions are but also so as to interrogate the rules that govern such forms of *problematization* altogether. The assetization of poverty certainly appears as a particularly problematic case (Mitchell 2008), but so does the assetization of natural resources (Mitchell 2011). What is the dominant problem in those cases, if not a proper revaluation of resources with the future in sight? Hence the analytical, empirical, and political emphasis on understanding the asset form and condition as the essential way forward. As scholars discuss whether or not the geological era known as the Anthropocene should be better dubbed Capitalocene (Moore 2015, 2016; Bonneuil and Fressoz 2016), research on assetization should start to ask the extent to which our current solutions are part of the problem.

References

Adkins, L., Cooper, M., and Konings, M. 2019. Class in the 21st Century: Asset Inflation and the New Logic of Inequality. *Environment and Planning A*, doi:10.1177/0308518X19873673.

Alliez, É., and Lazzarato, M. 2016. *Capital and Wars*. South Pasadena, CA: Semiotext(e).

Birch, K. 2017a. Rethinking *Value* in the Bio-Economy: Finance, Assetization and the Management of Value. *Science, Technology and Human Values* 42 (3): 460–490.

Birch, K. 2017b. *A Research Agenda for Neoliberalism*. Cheltenham: Edward Elgar.

Birch, K. 2018. What Is the Asset Condition? *European Journal of Sociology* 59 (3): 500–506.

Blyth, M. 2013. *Austerity: The History of a Dangerous Idea*. Oxford: Oxford University Press.

Bonneuil, C., and Fressoz J.-B. 2016. *The Shock of the Anthropocene: The Earth, History and Us*. London: Verso.

Bridge, G., Bulkeley, H., Langley, P., and van Veelen, B. 2019. Pluralizing and Problematizing Carbon Finance. *Progress in Human Geography*, doi:10.1177/0309132519856260.

Chiapello, E. 2015. Financialisation of Valuation. *Human Studies* 38 (1): 13–35.

Cooper, M. 2017. *Family Values: Between Neoliberalism and the New Social Conservatism*. New York: Zone Books.

Davies, W. 2014. *The Limits of Neoliberalism: Authority, Sovereignty and the Logic of Competition*. London: Sage.

Ducastel, A., and Anseeuw, W. 2017. Agriculture as an Asset Class: Reshaping the South African Farming Sector. *Agriculture and Human Values* 34 (1): 199–209.

Feher, M. 2009. Self-Appreciation; or, the Aspirations of Human Capital. *Public Culture* 21 (1): 21–41.

Feher, M. 2018. *Rated Agency: Investee Politics in a Speculative Age*. New York: Zone Books.

Geiger, S., and Gross, N. 2019. A Tidal Wave of Inevitable Data? Assetization in the Consumer Genomics Testing Industry. *Business & Society*, doi:10.1177/0007650319826307.

Godechot, O. 2017. *Wages, Bonuses and Appropriation of Profit in the Financial Industry: The Working Rich*. London: Routledge.

Haskel, J., and Westlake, S. 2018. *Capitalism without Capital: The Rise of the Intangible Economy*. Princeton, NJ: Princeton University Press.

Konings, M. 2018. *Capital and Time: For a New Critique of Neoliberal Reason*. Stanford, CA: Stanford University Press.

Larder, N., Sippel, S. R., and Argent, N. 2018. The Redefined Role of Finance in Australian Agriculture. *Australian Geographer* 49 (3): 397–418.

Leyshon, A., and Thrift, N. 2007. The Capitalization of Almost Everything: The Future of Finance and Capitalism. *Theory, Culture and Society* 24 (7–8): 97–115.

Mitchell, T. 2008. Rethinking Economy. *Geoforum* 39 (3): 1116–1121.

Mitchell, T. 2011. *Carbon Democracy: Political Power in the Age of Oil*. London: Verso.

Moore, J. W. 2015. *Capitalism in the Web of Life: Ecology and the Accumulation of Capital*. London: Verso.

Moore, J. W., ed. 2016. *Anthropocene or Capitalocene? Nature, History, and the Crisis of Capitalism*. Oakland, CA: PM Press.

Muniesa, F., Doganova, L., Ortiz, H., Pina-Stranger, A., Paterson, F., Bourgoin, A., Ehrenstein, V., Juven, P.-A., Pontille, D., Saraç-Lesavre, B., and Yon, G. 2017. *Capitalization: A Cultural Guide*. Paris: Presses des Mines.

Neyland, D. 2018. On the Transformation of Children at-Risk into an Investment Proposition: A Study of Social Impact Bonds as an Anti-market Device. *Sociological Review* 66 (3): 492–510.

Nieborg, D., and Poell, T. 2018. The Platformization of Cultural Production: Theorizing the Contingent Cultural Commodity. *new media & society* 20 (11): 4275–4292.

Nitzan, J., and Bichler, S. 2009. *Capital as Power: A Study of Order and Creorder*. London: Routledge.

O'Brien, P., O'Neill, P., and Pike, A. 2019. Funding, Financing and Governance Urban Infrastructure. *Urban Studies* 56 (7): 1291–1303.

OECD. 2012. *New Sources of Growth: Knowledge-Based Capital Driving Investment and Productivity in the 21st Century*. Paris: Organization for Economic Cooperation and Development.

Ouma, S. 2019. This Can('t) Be an Asset Class: The World of Money Management, "Society," and the Contested Morality of Farmland Investments. *Environment and Planning A: Economy and Space*, doi:10.1177/0308518X18790051.

Perzanowski, A., and Schultz, J. 2016. *The End of Ownership: Personal Property in the Digital Economy*. Cambridge, MA: MIT Press.

Pistor, K. 2019. *The Code of Capital: How the Law Creates Wealth and Inequality*. Princeton, NJ: Princeton University Press.

Sadowski, J. 2019. When Data Is Capital: Datafication, Accumulation, and Extraction. *Big Data & Society*, doi:10.1177/2053951718820549.

Streeck, W. 2014. *Buying Time: The Delayed Crisis of Democratic Capitalism*. London: Verso.

Tooze, A. 2018. *Crashed: How a Decade of Financial Crises Changed the World*. New York: Viking.

Tyfield, D. 2018. *Liberalism 2.0 and the Rise of China*. London: Routledge.

Ward, C., and Swyngedouw, E. 2018. Neoliberalisation from the Ground Up: Insurgent Capital, Regional Struggle, and the Assetisation of Land. *Antipode* 50 (4): 1077–1097.

West, S. M. 2019. Data Capitalism: Redefining the Logics of Surveillance and Privacy. *Business & Society* 58 (1): 20–41.

Zuboff, S. 2019. *The Age of Surveillance Capitalism*. New York: Public Affairs.

Contributors

Thomas Beauvisage is senior researcher at Orange Labs.

Kean Birch is associate professor at York University.

Veit Braun is a research associate in the Institute of Sociology at Goethe-University Frankfurt.

Natalia Buier is an associate at the Max Planck Institute for Social Anthropology.

Béatrice Cointe is a fellow at the University of Oslo TIK Centre for Technology, Innovation and Culture.

Paul Robert Gilbert is senior lecturer in international development at the University of Sussex.

Hyo Yoon Kang is senior lecturer in law at the University of Kent.

Les Levidow is senior research fellow at the Open University.

Kevin Mellet is senior researcher at Orange Labs.

Sveta Milyaeva is lecturer in sociology at the University of Bristol.

Fabian Muniesa is senior researcher at Mines ParisTech.

Alain Nadaï is senior research fellow at CIRED.

Daniel Neyland is professor of sociology at Goldsmiths, University of London.

Victor Roy is a research fellow at University College London's Institute for Innovation and Public Purpose and a physician at Boston Medical Center.

James W. Williams is associate professor at York University.

Index

Inside Technology Series

Edited by Wiebe E. Bijker, W. Bernard Carlson, and Trevor Pinch

Kean Birch and Fabian Muniesa, *Assetization:* Turning Things into Assets in Technoscientific Capitalism

David Demortain, *The Science of Bureaucracy: Risk Decision-Making and the US Environmental Protection Agency*

Joeri Bruynincx, *Listening in the Field: Recording and the Science of Birdsong*

Edward Jones-Imhotep, *The Unreliable Nation: Hostile Nature and Technological Failure in the Cold War*

Jennifer L. Lieberman, *Power Lines: Electricity in American Life and Letters, 1882–1952*

Jess Bier, *Mapping Israel, Mapping Palestine: Occupied Landscapes of International Technoscience*

Benoît Godin, *Models of Innovation: The History of an Idea*

Stephen Hilgartner, *Reordering Life: Knowledge and Control in the Genomics Revolution*

Brice Laurent, *Democratic Experiments: Problematizing Nanotechnology and Democracy in Europe and the United States*

Cyrus C. M. Mody, *The Long Arm of Moore's Law: Microelectronics and American Science*

Tiago Saraiva, *Fascist Pigs: Technoscientific Organisms and the History of Fascism*

Teun Zuiderent-Jerak, *Situated Interventions: Sociological Experiments in Healthcare*

Basile Zimmermann, *Technology and Cultural Difference: Electronic Music Devices, Social Networking Sites, and Computer Encodings in Contemporary China*

Andrew J. Nelson, *The Sound of Innovation: Stanford and the Computer Music Revolution*

Sonja D. Schmid, *Producing Power: The Pre-Chernobyl History of the Soviet Nuclear Industry*

Casey O'Donnell, *Developer's Dilemma: The Secret World of Videogame Creators*

Christina Dunbar-Hester, *Low Power to the People: Pirates, Protest, and Politics in FM Radio Activism*

Eden Medina, Ivan da Costa Marques, and Christina Holmes, editors, *Beyond Imported Magic: Essays on Science, Technology, and Society in Latin America*

Anique Hommels, Jessica Mesman, and Wiebe E. Bijker, editors, *Vulnerability in Technological Cultures: New Directions in Research and Governance*

Amit Prasad, *Imperial Technoscience: Transnational Histories of MRI in the United States, Britain, and India*

Charis Thompson, *Good Science: The Ethical Choreography of Stem Cell Research*

Tarleton Gillespie, Pablo J. Boczkowski, and Kirsten A. Foot, editors, *Media Technologies: Essays on Communication, Materiality, and Society*

Catelijne Coopmans, Janet Vertesi, Michael Lynch, and Steve Woolgar, editors, *Representation in Scientific Practice Revisited*

Rebecca Slayton, *Arguments that Count: Physics, Computing, and Missile Defense, 1949–2012*

Stathis Arapostathis and Graeme Gooday, *Patently Contestable: Electrical Technologies and Inventor Identities on Trial in Britain*

Jens Lachmund, *Greening Berlin: The Co-Production of Science, Politics, and Urban Nature*

Chikako Takeshita, *The Global Biopolitics of the IUD: How Science Constructs Contraceptive Users and Women's Bodies*

Cyrus C. M. Mody, *Instrumental Community: Probe Microscopy and the Path to Nanotechnology*

Morana Alač, *Handling Digital Brains: A Laboratory Study of Multimodal Semiotic Interaction in the Age of Computers*

Gabrielle Hecht, editor, *Entangled Geographies: Empire and Technopolitics in the Global Cold War*

Michael E. Gorman, editor, *Trading Zones and Interactional Expertise: Creating New Kinds of Collaboration*

Matthias Gross, *Ignorance and Surprise: Science, Society, and Ecological Design*

Andrew Feenberg, *Between Reason and Experience: Essays in Technology and Modernity*

Wiebe E. Bijker, Roland Bal, and Ruud Hendricks, *The Paradox of Scientific Authority: The Role of Scientific Advice in Democracies*

Park Doing, *Velvet Revolution at the Synchrotron: Biology, Physics, and Change in Science*

Gabrielle Hecht, *The Radiance of France: Nuclear Power and National Identity after World War II*

Richard Rottenburg, *Far-Fetched Facts: A Parable of Development Aid*

Michel Callon, Pierre Lascoumes, and Yannick Barthe, *Acting in an Uncertain World: An Essay on Technical Democracy*

Ruth Oldenziel and Karin Zachmann, editors, *Cold War Kitchen: Americanization, Technology, and European Users*

Deborah G. Johnson and Jameson W. Wetmore, editors, *Technology and Society: Building Our Sociotechnical Future*

Trevor Pinch and Richard Swedberg, editors, *Living in a Material World: Economic Sociology Meets Science and Technology Studies*

Christopher R. Henke, *Cultivating Science, Harvesting Power: Science and Industrial Agriculture in California*

Helga Nowotny, *Insatiable Curiosity: Innovation in a Fragile Future*

Karin Bijsterveld, *Mechanical Sound: Technology, Culture, and Public Problems of Noise in the Twentieth Century*

Peter D. Norton, *Fighting Traffic: The Dawn of the Motor Age in the American City*

Joshua M. Greenberg, *From Betamax to Blockbuster: Video Stores tand the Invention of Movies on Video*

Mikael Hård and Thomas J. Misa, editors, *Urban Machinery: Inside Modern European Cities*

Christine Hine, *Systematics as Cyberscience: Computers, Change, and Continuity in Science*

Wesley Shrum, Joel Genuth, and Ivan Chompalov, *Structures of Scientific Collaboration*

Shobita Parthasarathy, *Building Genetic Medicine: Breast Cancer, Technology, and the Comparative Politics of Health Care*

Kristen Haring, *Ham Radio's Technical Culture*

Atsushi Akera, *Calculating a Natural World: Scientists, Engineers and Computers during the Rise of U.S. Cold War Research*

Donald MacKenzie, *An Engine, Not a Camera: How Financial Models Shape Markets*

Geoffrey C. Bowker, *Memory Practices in the Sciences*

Christophe Lécuyer, *Making Silicon Valley: Innovation and the Growth of High Tech, 1930–1970*

Anique Hommels, *Unbuilding Cities: Obduracy in Urban Sociotechnical Change*

David Kaiser, editor, *Pedagogy and the Practice of Science: Historical and Contemporary Perspectives*

Charis Thompson, *Making Parents: The Ontological Choreography of Reproductive Technology*

Pablo J. Boczkowski, *Digitizing the News: Innovation in Online Newspapers*

Dominique Vinck, editor, *Everyday Engineering: An Ethnography of Design and Innovation*

Nelly Oudshoorn and Trevor Pinch, editors, *How Users Matter: The Co-Construction of Users and Technology*

Peter Keating and Alberto Cambrosio, *Biomedical Platforms: Realigning the Normal and the Pathological in Late-Twentieth-Century Medicine*

Paul Rosen, *Framing Production: Technology, Culture, and Change in the British Bicycle Industry*

Maggie Mort, *Building the Trident Network: A Study of the Enrollment of People, Knowledge, and Machines*

Donald MacKenzie, *Mechanizing Proof: Computing, Risk, and Trust*

Geoffrey C. Bowker and Susan Leigh Star, *Sorting Things Out: Classification and Its Consequences*

Charles Bazerman, *The Languages of Edison's Light*

Janet Abbate, *Inventing the Internet*

Herbert Gottweis, *Governing Molecules: The Discursive Politics of Genetic Engineering in Europe and the United States*

Kathryn Henderson, *On Line and On Paper: Visual Representation, Visual Culture, and Computer Graphics in Design Engineering*

Susanne K. Schmidt and Raymund Werle, *Coordinating Technology: Studies in the International Standardization of Telecommunications*

Marc Berg, *Rationalizing Medical Work: Decision Support Techniques and Medical Practices*

Eda Kranakis, *Constructing a Bridge: An Exploration of Engineering Culture, Design, and Research in Nineteenth-Century France and America*

Paul N. Edwards, *The Closed World: Computers and the Politics of Discourse in Cold War America*

Donald MacKenzie, *Knowing Machines: Essays on Technical Change*

Wiebe E. Bijker, *Of Bicycles, Bakelites, and Bulbs: Toward a Theory of Sociotechnical Change*

Louis L. Bucciarelli, *Designing Engineers*

Geoffrey C. Bowker, *Science on the Run: Information Management and Industrial Geophysics at Schlumberger, 1920–1940*

Wiebe E. Bijker and John Law, editors, *Shaping Technology / Building Society: Studies in Sociotechnical Change*

Stuart Blume, *Insight and Industry: On the Dynamics of Technological Change in Medicine*

Donald MacKenzie, *Inventing Accuracy: A Historical Sociology of Nuclear Missile Guidance*

Pamela E. Mack, *Viewing the Earth: The Social Construction of the Landsat Satellite System*

H. M. Collins, *Artificial Experts: Social Knowledge and Intelligent Machines*

Lukas Engelmann and Christos Lynteris, *Sulphuric Utopias: The History of Maritime Fumigation*

Zara Mirmalek, *Making Time on Mars*
http://mitpress.mit.edu/books/series/inside-technology

Printed by Libri Plureos GmbH in Hamburg,
Germany